BAYESIAN STATISTICS THE FUN WAY

BAYESIAN STATISTICS THE FUN WAY

Understanding Statistics and Probability with Star Wars®, LEGO®, and Rubber Ducks

by Will Kurt

no starch press

San Francisco

Printed in USA

Second printing

23 22 21 20 19   2 3 4 5 6 7 8 9

ISBN-10: 1-59327-956-6
ISBN-13: 978-1-59327-956-1

Publisher: William Pollock
Production Editor: Laurel Chun
Cover Illustration: Josh Ellingson
Interior Design: Octopod Studios
Developmental Editor: Liz Chadwick
Technical Reviewer: Chelsea Parlett-Pelleriti
Copyeditor: Rachel Monaghan
Compositor: Danielle Foster
Proofreader: James Fraleigh
Indexer: Erica Orloff

For information on distribution, translations, or bulk sales, please contact No Starch Press, Inc. directly:
No Starch Press, Inc.
245 8th Street, San Francisco, CA 94103
phone: 1.415.863.9900; sales@nostarch.com
www.nostarch.com

Library of Congress Cataloging-in-Publication Data

Names: Kurt, Will, author.
Title: Bayesian statistics the fun way / Will Kurt.
Description: San Francisco : No Starch Press, Inc., [2019] | Includes
    index. | Summary: "An introduction to Bayesian statistics with simple
    and pop culture-based explanations. Topics covered include measuring
    your own uncertainty in a belief, applying Bayes' theorem, and
    calculating distributions"-- Provided by publisher.
Identifiers: LCCN 2019020743 (print) | LCCN 2019022412 (ebook) | ISBN
    9781593279561 (print) | ISBN 1593279566 (print)
Subjects: LCSH: Bayesian statistical decision theory. | Probabilities.
Classification: LCC QA279.5 .K87 2019 (print) | LCC QA279.5 (ebook) | DDC
    519.5/42--dc23
LC record available at https://lccn.loc.gov/2019020743

To Melanie, who reawoke in me a passion for words

About the Author

Will Kurt currently works as a data scientist at Wayfair, and has been using Bayesian statistics to solve real business problems for over half a decade. He frequently blogs about probability on his website, *CountBayesie.com*. Kurt is the author of *Get Programming with Haskell* (Manning Publications) and lives in Boston, Massachusetts.

About the Technical Reviewer

Chelsea Parlett-Pelleriti is a PhD student in Computational and Data Science, and has a long-standing love of all things lighthearted and statistical. She is also a freelance statistics writer, contributing to projects including the YouTube series *Crash Course Statistics* and The Princeton Review's *Cracking the AP Statistics Exam*. She currently lives in Southern California.

BRIEF CONTENTS

CONTENTS IN DETAIL

7
BAYES' THEOREM WITH LEGO 67

8
THE PRIOR, LIKELIHOOD, AND POSTERIOR OF BAYES' THEOREM 73

9
BAYESIAN PRIORS AND WORKING WITH
PROBABILITY DISTRIBUTIONS 83

PART III: PARAMETER ESTIMATION

10
INTRODUCTION TO AVERAGING AND PARAMETER ESTIMATION 93

PART IV: HYPOTHESIS TESTING: THE HEART OF STATISTICS

15
FROM PARAMETER ESTIMATION TO HYPOTHESIS TESTING: BUILDING A BAYESIAN A/B TEST   149

16
INTRODUCTION TO THE BAYES FACTOR AND POSTERIOR ODDS: THE COMPETITION OF IDEAS   157

17
BAYESIAN REASONING IN THE TWILIGHT ZONE   167

18
WHEN DATA DOESN'T CONVINCE YOU   175

ACKNOWLEDGMENTS

Writing a book is really an incredible effort that involves the hard work of many people. Even with all the names following I can only touch on some of the many people that have made this book possible. I would like to start by thanking my son, Archer, for always keeping me curious and inspiring me.

The books published by No Starch have long been my some of my favorite books to read and it is a real honor to get to work with the amazing team there to produce this book. I give tremendous thanks to my editors, reviewers, and the incredible team at No Starch. Liz Chadwick originally approached me about creating this book and provided excellent editiorial feedback and guidance through the entire porcess of this book. Laurel Chun made sure the entire process of going from some messy R notebooks to a full fledged book went incredibly smoothly. Chelsea Parlett-Pelleriti went well beyond the requirements of a technical reviewer and really helped to make this book the best it can be. Frances Saux added many insightful comments to the later chapters of the book. And of course thank you to Bill Pollock for creating such a delightful publishing company.

As an English literature major in undergrad I never could have imagined writing a book on any mathematical subject. There are a few people who were really essential to helping me see the wonder of mathematics. I will forever be grateful to my college roommate, Greg Muller, who showed a crazy English major just how exciting and interesting the world of mathematics can be. Professor Anatoly Temkin at Boston University opened the doors to mathematical thinking for me by teaching me to always answer the question, "what does this mean?" And of course a huge thanks to Richard Kelley who, when I found myself in the desert for many years, provided an oasis of mathematical conversations and guidence. I would also like to give a shoutout to the data science team at Bombora, especially Patrick Kelley, who provided so many wonderful questions and coversations, some of which found their way into this book. I will also be forever grateful to the readers of my blog, *Count Bayesie*, who have always provided wonderful questions and insights. Among these readers, I would especially like to thank the commentor Nevin who helped correct some early misunderstandings I had.

Finally I want to give thanks to some truly great authors in Bayesian statistics whose books have done a great deal to guide my own growth in the subject. John Kruschke's *Doing Bayesian Data Analysis* and *Bayesian Data Analysis* by Andrew Gelman, et al. are great books everyone should read. By far the most influential book on my own thinking is E.T. Jaynes' phenomenal *Probability Theory: The Logic of Science*, and I'd like to add thanks to Aubrey Clayton for making a series of lectures on this challenging book which really helped clarify it for me.

INTRODUCTION

Virtually everything in life is, to some extent, uncertain. This may seem like a bit of an exaggeration, but to see the truth of it you can try a quick experiment. At the start of the day, write down something you think will happen in the next half-hour, hour, three hours, and six hours. Then see how many of these things happen exactly like you imagined. You'll quickly realize that your day is full of uncertainties. Even something as predictable as "I will brush my teeth" or "I'll have a cup of coffee" may not, for some reason or another, happen as you expect.

For most of the uncertainties in life, we're able to get by quite well by planning our day. For example, even though traffic might make your morning commute longer than usual, you can make a pretty good estimate about what time you need to leave home in order to get to work on time. If you have a super-important morning meeting, you might leave earlier to allow for delays. We all have an innate sense of how to deal with uncertain situations and reason about uncertainty. When you think this way, you're starting to think *probabilistically*.

Why Learn Statistics?

The subject of this book, Bayesian statistics, helps us get better at reasoning about uncertainty, just as studying logic in school helps us to see the errors in everyday logical thinking. Given that virtually everyone deals with uncertainty in their daily life, as we just discussed, this makes the audience for this book pretty wide. Data scientists and researchers already using statistics will benefit from a deeper understanding and intuition for how these tools work. Engineers and programmers will learn a lot about how they can better quantify decisions they have to make (I've even used Bayesian analysis to identify causes of software bugs!). Marketers and salespeople can apply the ideas in this book when running A/B tests, trying to understand their audience, and better assessing the value of opportunities. Anyone making high-level decisions should have at least a basic sense of probability so they can make quick back-of-the-envelope estimates about the costs and benefits of uncertain decisions. I wanted this book to be something a CEO could study on a flight and develop a solid enough foundation by the time they land to better assess choices that involve probabilities and uncertainty.

I honestly believe that everyone will benefit from thinking about problems in a Bayesian way. With Bayesian statistics, you can use mathematics to model that uncertainty so you can make better choices given limited information. For example, suppose you need to be on time for work for a particularly important meeting and there are two different routes you could take. The first route is usually faster, but has pretty regular traffic back-ups that can cause huge delays. The second route takes longer in general but is less prone to traffic. Which route should you take? What type of information would you need to decide this? And how certain can you be in your choice? Even just a small amount of added complexity requires some extra thought and technique.

Typically when people think of statistics, they think of scientists working on a new drug, economists following trends in the market, analysts predicting the next election, baseball managers trying to build the best team with fancy math, and so on. While all of these are certainly fascinating uses of statistics, understanding the basics of Bayesian reasoning can help you in far more areas in everyday life. If you've ever questioned some new finding reported in the news, stayed up late browsing the web wondering if you have a rare disease, or argued with a relative over their irrational beliefs about the world, learning Bayesian statistics will help you reason better.

What Is "Bayesian" Statistics?

You may be wondering what all this "Bayesian" stuff is. If you've ever taken a statistics class, it was likely based on *frequentist statistics*. Frequentist statistics is founded on the idea that probability represents the frequency with which something happens. If the probability of getting heads in a single coin toss is 0.5, that means after a single coin toss we can expect to get one-half of a head of a coin (with two tosses we can expect to get one head, which makes more sense).

Bayesian statistics, on the other hand, is concerned with how probabilities represent how uncertain we are about a piece of information. In Bayesian terms, if the probability of getting heads in a coin toss is 0.5, that means we are equally unsure about whether we'll get heads or tails. For problems like coin tosses, both frequentist and Bayesian approaches seem reasonable, but when you're quantifying your belief that your favorite candidate will win the next election, the Bayesian interpretation makes much more sense. After all, there's only one election, so speaking about how frequently your favorite candidate will win doesn't make much sense. When doing Bayesian statistics, we're just trying to accurately describe what we believe about the world given the information we have.

One particularly nice thing about Bayesian statistics is that, because we can view it simply as reasoning about uncertain things, all of the tools and techniques of Bayesian statistics make intuitive sense.

Bayesian statistics is about looking at a problem you face, figuring out how you want to describe it mathematically, and then using reason to solve it. There are no mysterious tests that give results that you aren't quite sure of, no distributions you have to memorize, and no traditional experiment designs you must perfectly replicate. Whether you want to figure out the probability that a new web page design will bring you more customers, if your favorite sports team will win the next game, or if we really are alone in the universe, Bayesian statistics will allow you to start reasoning about these things mathematically using just a few simple rules and a new way of looking at problems.

What's in This Book

Here's a quick breakdown of what you'll find in this book.

Part I: Introduction to Probability

Chapter 1: Bayesian Thinking and Everyday Reasoning   This first chapter introduces you to Bayesian thinking and shows you how similar it is to everyday methods of thinking critically about a situation. We'll explore the probability that a bright light outside your window at night is a UFO based on what you already know and believe about the world.

Chapter 2: Measuring Uncertainty   In this chapter you'll use coin toss examples to assign actual values to your uncertainty in the form of probabilities: a number from 0 to 1 that represents how certain you are in your belief about something.

Chapter 3: The Logic of Uncertainty   In logic we use AND, NOT, and OR operators to combine true or false facts. It turns out that probability has similar notions of these operators. We'll investigate how to reason about the best mode of transport to get to an appointment, and the chances of you getting a traffic ticket.

Chapter 4: Creating a Binomial Probability Distribution   Using the rules of probability as logic, in this chapter, you'll build your own probability distribution, the binomial distribution, which you can apply to many probability problems that share a similar structure. You'll try to predict the probability of getting a specific famous statistician collectable card in a Gacha card game.

Chapter 5: The Beta Distribution   Here you'll learn about your first continuous probability distribution and get an introduction to what makes statistics different from probability. The practice of statistics involves trying to figure out what unknown probabilities might be based on data. In this chapter's example, we'll investigate a mysterious coin-dispensing box and the chances of making more money than you lose.

Part II: Bayesian Probability and Prior Probabilities

Chapter 6: Conditional Probability   In this chapter, you'll condition probabilities based on your existing information. For example, knowing whether someone is male or female tells us how likely they are to be color blind. You'll also be introduced to Bayes' theorem, which allows us to reverse conditional probabilities.

Chapter 7: Bayes' Theorem with LEGO   Here you'll gain a better intuition for Bayes' theorem by reasoning about LEGO bricks! This chapter will give you a spatial sense of what Bayes' theorem is doing mathematically.

Chapter 8: The Prior, Likelihood, and Posterior of Bayes' Theorem   Bayes' theorem is typically broken into three parts, each of which performs its own function in Bayesian reasoning. In this chapter, you'll learn what they're called and how to use them by investigating whether an apparent break-in was really a crime or just a series of coincidences.

Chapter 9: Bayesian Priors and Working with Probability Distributions   This chapter explores how we can use Bayes' theorem to better understand the classic asteroid scene from *Star Wars: The Empire Strikes Back*, through which you'll gain a stronger understanding of prior probabilities in Bayesian statistics. You'll also see how you can use entire distributions as your prior.

Part III: Parameter Estimation

Chapter 10: Introduction to Averaging and Parameter Estimation
Parameter estimation is the method we use to formulate a best guess
for an uncertain value. The most basic tool in parameter estimation is
to simply average your observations. In this chapter we'll see why this
works by analyzing snowfall levels.

Chapter 11: Measuring the Spread of Our Data   Finding the mean
is a useful first step in estimating parameters, but we also need a way
to account for how spread out our observations are. Here you'll be
introduced to mean absolute deviation (MAD), variance, and standard
deviation as ways to measure how spread out our observations are.

Chapter 12: The Normal Distribution   By combining our mean and
standard deviation, we get a very useful distribution for making estimates: the normal distribution. In this chapter, you'll learn how to use
the normal distribution to not only estimate unknown values but also
to know how certain you are in those estimates. You'll use these new
skills to time your escape during a bank heist.

**Chapter 13: Tools of Parameter Estimation: The PDF, CDF, and
Quantile Function**   Here you'll learn about the PDF, CDF, and quantile function to better understand the parameter estimations you're
making. You'll estimate email conversion rates using these tools and see
what insights each provides.

Chapter 14: Parameter Estimation with Prior Probabilities
The best way to improve our parameter estimates is to include a prior
probability. In this chapter, you'll see how adding prior information
about the past success of email click-through rates can help us better
estimate the true conversion rate for a new email.

**Chapter 15: From Parameter Estimation to Hypothesis Testing:
Building a Bayesian A/B Test**   Now that we can estimate uncertain
values, we need a way to compare two uncertain values in order to test
a hypothesis. You'll create an A/B test to determine how confident you
are in a new method of email marketing.

Part IV: Hypothesis Testing: The Heart of Statistics

**Chapter 16: Introduction to the Bayes Factor and Posterior Odds: The
Competition of Ideas**   Ever stay up late, browsing the web, wondering if you might have a super-rare disease? This chapter will introduce
another approach to testing ideas that will help you determine how
worried you should actually be!

Chapter 17: Bayesian Reasoning in The Twilight Zone   How much do
you believe in psychic powers? In this chapter, you'll develop your own
mind-reading skills by analyzing a situation from a classic episode of
The Twilight Zone.

Chapter 18: When Data Doesn't Convince You Sometimes data doesn't seem to be enough to change someone's mind about a belief or help you win an argument. Learn how you can change a friend's mind about something you disagree on and why it's not worth your time to argue with your belligerent uncle!

Chapter 19: From Hypothesis Testing to Parameter Estimation
Here we come full circle back to parameter estimation by looking at how to compare a range of hypotheses. You'll derive your first example of statistics, the beta distribution, using the tools that we've covered for simple hypothesis tests to analyze the fairness of a particular fairground game.

Appendix A: A Quick Introduction to R This quick appendix will teach you the basics of the R programming language.

Appendix B: Enough Calculus to Get By Here we'll cover just enough calculus to get you comfortable with the math used in the book.

Background for Reading the Book

The only requirement of this book is basic high school algebra. If you flip forward, you'll see a few instances of math, but nothing particularly onerous. We'll be using a bit of code written in the R programming language, which I'll provide and talk through, so there's no need to have learned R beforehand. We'll also touch on calculus, but again no prior experience is required, and the appendixes will give you enough information to cover what you'll need.

In other words, this book aims to help you start thinking about problems in a mathematical way without requiring significant mathematical background. When you finish reading it, you may find yourself inadvertently writing down equations to describe problems you see in everyday life!

If you do happen to have a strong background in statistics (even Bayesian statistics), I believe you'll still have a fun time reading through this book. I have always found that the best way to understand a field well is to revisit the fundamentals over and over again, each time in a different light. Even as the author of this book, I found plenty of things that surprised me just in the course of the writing process!

Now Off on Your Adventure!

As you'll soon see, aside from being very useful, Bayesian statistics can be a lot of fun! To help you learn Bayesian reasoning we'll be taking a look at LEGO bricks, *The Twilight Zone*, *Star Wars*, and more. You'll find that once you begin thinking probabilistically about problems, you'll start using Bayesian statistics all over the place. This book is designed to be a pretty quick and enjoyable read, so turn the page and let's begin our adventure in Bayesian statistics!

PART I

INTRODUCTION TO PROBABILITY

1

BAYESIAN THINKING AND EVERYDAY REASONING

In this first chapter, I'll give you an overview of *Bayesian reasoning*, the formal process we use to update our beliefs about the world once we've observed some data. We'll work through a scenario and explore how we can map our everyday experience to Bayesian reasoning.

The good news is that you were already a Bayesian even before you picked up this book! Bayesian statistics is closely aligned with how people naturally use evidence to create new beliefs and reason about everyday problems; the tricky part is breaking down this natural thought process into a rigorous, mathematical one.

In statistics, we use particular calculations and models to more accurately quantify probability. For now, though, we won't use any math or models; we'll just get you familiar with the basic concepts and use our intuition to determine probabilities. Then, in the next chapter, we'll put exact numbers to probabilities. Throughout the rest of the book, you'll learn how we can use rigorous mathematical techniques to formally model and reason about the concepts we'll cover in this chapter.

Reasoning About Strange Experiences

> One night you are suddenly awakened by a bright light at your window. You jump up from bed and look out to see a large object in the sky that can only be described as saucer shaped. You are generally a skeptic and have never believed in alien encounters, but, completely perplexed by the scene outside, you find yourself thinking, *Could this be a UFO?!*

Bayesian reasoning involves stepping through your thought process when you're confronted with a situation to recognize when you're making probabilistic assumptions, and then using those assumptions to update your beliefs about the world. In the UFO scenario, you've already gone through a full Bayesian analysis because you:

1. Observed data
2. Formed a hypothesis
3. Updated your beliefs based on the data

This reasoning tends to happen so quickly that you don't have any time to analyze your own thinking. You created a new belief without questioning it: whereas before you did not believe in the existence of UFOs, after the event you've updated your beliefs and now think you've seen a UFO.

In this chapter, you'll focus on structuring your beliefs and the process of creating them so you can examine it more formally, and we'll look at quantifying this process in chapters to come.

Let's look at each step of reasoning in turn, starting with observing data.

Observing Data

Founding your beliefs on data is a key component of Bayesian reasoning. Before you can draw any conclusions about the scene (such as claiming what you see is a UFO), you need to understand the data you're observing, in this case:

- An extremely bright light outside your window
- A saucer-shaped object hovering in the air

Based on your past experience, you would describe what you saw out your window as "surprising." In probabilistic terms, we could write this as:

$$P(\text{bright light outside window, saucer-shaped object in sky}) = \text{very low}$$

where P denotes *probability* and the two pieces of data are listed inside the parentheses. You would read this equation as: "The probability of observing bright lights outside the window and a saucer-shaped object in the sky is very low." In probability theory, we use a comma to separate events when we're looking at the combined probability of multiple events. Note that this data does not contain anything specific about UFOs; it's simply made up of your observations—this will be important later.

We can also examine probabilities of single events, which would be written as:

$$P(\text{rain}) = \text{likely}$$

This equation is read as: "The probability of rain is likely."

For our UFO scenario, we're determining the probability of *both* events occurring *together*. The probability of one of these two events occurring on its own would be entirely different. For example, the bright lights alone could easily be a passing car, so on its own the probability of this event is more likely than its probability coupled with seeing a saucer-shaped object (and the saucer-shaped object would still be surprising even on its own).

So how are we determining this probability? Right now we're using our intuition—that is, our general sense of the likelihood of perceiving these events. In the next chapter, we'll see how we can come up with exact numbers for our probabilities.

Holding Prior Beliefs and Conditioning Probabilities

You are able to wake up in the morning, make your coffee, and drive to work without doing a lot of analysis because you hold *prior beliefs* about how the world works. Our prior beliefs are collections of beliefs we've built up over a lifetime of experiences (that is, of observing data). You believe that the sun will rise because the sun has risen every day since you were born. Likewise, you might have a prior belief that when the light is red for oncoming traffic at an intersection, and your light is green, it's safe to drive through the intersection. Without prior beliefs, we would go to bed terrified each night that the sun might not rise tomorrow, and stop at every intersection to carefully inspect oncoming traffic.

Our prior beliefs say that seeing bright lights outside the window at the same time as seeing a saucer-shaped object is a rare occurrence on Earth. However, if you lived on a distant planet populated by vast numbers of flying saucers, with frequent interstellar visitors, the probability of seeing lights and saucer-shaped objects in the sky would be much higher.

In a formula we enter prior beliefs after our data, separated with a |
like so:

$$P\left(\begin{array}{c}\text{bright light outside window, saucer-shaped}\\ \text{object in sky} \mid \text{experience on Earth}\end{array}\right) = \text{very low}$$

We would read this equation as: "The probability of observing bright lights and a saucer-shaped object in the sky, *given* our experience on Earth, is very low."

The probability outcome is called a *conditional probability* because we are *conditioning* the probability of one event occurring on the existence of something else. In this case, we're conditioning the probability of our observation on our prior experience.

In the same way we used P for probability, we typically use shorter variable names for events and conditions. If you're unfamiliar with reading equations, they can seem too terse at first. After a while, though, you'll find that shorter variable names aid readability and help you to see how equations generalize to larger classes of problems. We'll assign all of our data to a single variable, D:

$$D = \text{bright light outside window, saucer-shaped object in sky}$$

So from now on when we refer to the probability of set of data, we'll simply say, $P(D)$.

Likewise, we use the variable X to represent our prior belief, like so:

$$X = \text{experience on Earth}$$

We can now write this equation as $P(D \mid X)$. This is much easier to write and doesn't change the meaning.

Conditioning on Multiple Beliefs

We can add more than one piece of prior knowledge, too, if more than one variable is going to significantly affect the probability. Suppose that it's July 4th and you live in the United States. From prior experience you know that fireworks are common on the Fourth of July. Given your experience on Earth *and* the fact that it's July 4th, the probability of seeing lights in the sky is less unlikely, and even the saucer-shaped object could be related to some fireworks display. You could rewrite this equation as:

$$P\left(\begin{array}{l}\text{bright light outside window, saucer-shaped} \\ \text{object in sky} \mid \text{July 4th, experience on Earth}\end{array}\right) = \text{low}$$

Taking both these experiences into account, our conditional probability changed from "very low" to "low."

Assuming Prior Beliefs in Practice

In statistics, we don't usually explicitly include a condition for all of our existing experiences, because it can be assumed. For that reason, in this book we won't include a separate variable for this condition. However, in Bayesian analysis, it's essential to keep in mind that our understanding of the world is always conditioned on our prior experience in the world. For the rest of this chapter, we'll keep the "experience on Earth" variable around to remind us of this.

Forming a Hypothesis

So far we have our data, D (that we have seen a bright light and a saucer-shaped object), and our prior experience, X. In order to explain what you saw, you need to form some kind of *hypothesis*—a model about how the world works that makes a prediction. Hypotheses can come in many forms. All of our basic beliefs about the world are hypotheses:

- If you believe the Earth rotates, you predict the sun will rise and set at certain times.
- If you believe that your favorite baseball team is the best, you predict they will win more than the other teams.
- If you believe in astrology, you predict that the alignment of the stars will describe people and events.

Hypotheses can also be more formal or sophisticated:

- A scientist may hypothesize that a certain treatment will slow the growth of cancer.
- A quantitative analyst in finance may have a model of how the market will behave.
- A deep neural network may predict which images are animals and which ones are plants.

All of these examples are hypotheses because they have some way of understanding the world and use that understanding to make a prediction about how the world will behave. When we think of hypotheses in Bayesian statistics, we are usually concerned with how well they predict the data we observe.

When you see the evidence and think *A UFO!*, you are forming a hypothesis. The UFO hypothesis is likely based on countless movies and television shows you've seen in your prior experience. We would define our first hypothesis as:

$$H_1 = \text{A UFO is in my back yard!}$$

But what is this hypothesis predicting? If we think of this situation backward, we might ask, "If there was a UFO in your back yard, what would you expect to see?" And you might answer, "Bright lights and a saucer-shaped object." Because H_1 predicts the data D, when we observe our data given our hypothesis, the probability of the data increases. Formally we write this as:

$$P(D \mid H_1, X) \gg P(D \mid X)$$

This equation says: "The probability of seeing bright lights and a saucer-shaped object in the sky, given my belief that this is a UFO and my prior experience, is much higher [indicated by the double greater-than sign >>] than just seeing bright lights and a saucer-shaped object in the sky without explanation." Here we've used the language of probability to demonstrate that our hypothesis explains the data.

Spotting Hypotheses in Everyday Speech

It's easy to see a relationship between our everyday language and probability. Saying something is "surprising," for example, might be the same as saying it has low-probability data based on our prior experiences. Saying something "makes sense" might indicate we have high-probability data based on our prior experiences. This may seem obvious once pointed out, but the key to probabilistic reasoning is to think carefully about how you interpret data, create hypotheses, and change your beliefs, even in an ordinary, everyday scenario. Without H_1, you'd be in a state of confusion because you have no explanation for the data you observed.

Gathering More Evidence and Updating Your Beliefs

Now you have your data and a hypothesis. However, given your prior experience as a skeptic, that hypothesis still seems pretty outlandish. In order to improve your state of knowledge and draw more reliable conclusions, you need to collect more data. This is the next step in statistical reasoning, as well as in your own intuitive thinking.

To collect more data, we need to make more observations. In our scenario, you look out your window to see what you can observe:

> As you look toward the bright light outside, you notice more
> lights in the area. You also see that the large saucer-shaped object
> is held up by wires, and notice a camera crew. You hear a loud
> clap and someone call out "Cut!"

You have, very likely, instantly changed your mind about what you think is happening in this scene. Your inference before was that you might be witnessing a UFO. Now, with this new evidence, you realize it looks more like someone is shooting a movie nearby.

With this thought process, your brain has once again performed some sophisticated Bayesian analysis in an instant! Let's break down what happened in your head in order to reason about events more carefully.

You started with your initial hypothesis:

$$H_1 = \text{A UFO has landed!}$$

In isolation, this hypothesis, given your experience, is extremely unlikely:

$$P\left(H_1 \mid X\right) = \text{very, very low}$$

However, it was the only useful explanation you could come up with given the data you had available. When you observed additional data, you immediately realized that there's another possible hypothesis—that a movie is being filmed nearby:

$$H_2 = \text{A film is being made outside your window}$$

In isolation, the probability of this hypothesis is also intuitively very low (unless you happen to live near a movie studio):

$$P(H_2 \mid X) = \text{very low}$$

Notice that we set the probability of H_1 as "very, very low" and the probability of H_2 as just "very low." This corresponds to your intuition: if someone came up to you, without any data, and asked, "Which do you think is more likely, a UFO appearing at night in your neighborhood or a movie being filmed next door?" you would say the movie scenario is more likely than a UFO appearance.

Now we just need a way to take our new data into account when changing our beliefs.

Comparing Hypotheses

You first accepted the UFO hypothesis, despite it being unlikely, because you didn't initially have any other explanation. Now, however, there's another possible explanation—a movie being filmed—so you have formed an *alternate hypothesis*. Considering alternate hypotheses is the process of comparing multiple theories using the data you have.

When you see the wires, film crew, and additional lights, your data changes. Your updated data are:

$$D_{\text{updated}} = \text{bright lights, saucer-shaped object,}$$

$$\text{wires, film crew, other lights, etc.} \ldots$$

On observing this extra data, you change your conclusion about what was happening. Let's break this process down into Bayesian reasoning. Your first hypothesis, H_1, gave you a way to explain your data and end your confusion, but with your additional observations H_1 no longer explains the data well. We can write this as:

$$P(D_{\text{updated}} \mid H_1, X) = \text{very, very low}$$

You now have a new hypothesis, H_2, which explains the data much better, written as follows:

$$P(D_{\text{updated}} \mid H_2, X) \gg P(D_{\text{updated}} \mid H_1, X)$$

The key here is to understand that we're comparing how well each of these hypotheses explains the observed data. When we say, "The probability of the data, given the second hypothesis, is much greater than the first," we're saying that what we observed is better explained by the second hypothesis. This brings us to the true heart of Bayesian analysis: *the test of your beliefs is how well they explain the world.* We say that one belief is more accurate than another because it provides a better explanation of the world we observe.

Mathematically, we express this idea as the ratio of the two probabilities:

$$\frac{P\left(D_{\text{updated}} \mid H_2, X\right)}{P\left(D_{\text{updated}} \mid H_1, X\right)}$$

When this ratio is a large number, say 1,000, it means "H_2 explains the data 1,000 times better than H_1." Because H_2 explains the data many times better than another H_1, we update our beliefs from H_1 to H_2. This is exactly what happened when you changed your mind about the likely explanation for what you observed. You now believe that what you've seen is a movie being made outside your window, because this is a more likely explanation of all the data you observed.

Data Informs Belief; Belief Should Not Inform Data

One final point worth stressing is that the only absolute in all these examples is your data. Your hypotheses change, and your experience in the world, X, may be different from someone else's, but the data, D, is shared by all.

Consider the following two formulas. The first is one we've used throughout this chapter:

$$P(D \mid H, X)$$

which we read as "The probability of the data given my hypotheses and experience in the world," or more plainly, "How well my beliefs explain what I observe."

But there is a reversal of this, common in everyday thinking, which is:

$$P(H \mid D, X)$$

We read this as "The probability of *my beliefs* given the data and my experiences in the world," or "How well what I observe supports what I believe."

In the first case, we change our beliefs according to data we gather and observations we make about the world that describe it better. In the second case, we gather data to support our existing beliefs. Bayesian thinking is about changing your mind and updating how you understand the world. The data we observe is all that is real, so our beliefs ultimately need to shift until they align with the data.

In life, too, your beliefs should always be mutable.

> As the film crew packs up, you notice that all the vans bear military insignia. The crew takes off their coats to reveal army fatigues and you overhear someone say, "Well, that should have fooled anyone who saw that . . . good thinking."

With this new evidence, your beliefs may shift again!

Wrapping Up

Let's recap what you've learned. Your beliefs start with your existing experience of the world, X. When you observe data, D, it either aligns with your experience, $P(D|X)$ = very high, or it surprises you, $P(D|X)$ = very low. To understand the world, you rely on beliefs you have about what you observe, or hypotheses, H. Oftentimes a new hypothesis can help you explain the data that surprises you, $P(D|H, X) >> P(D|X)$. When you gather new data or come up with new ideas, you can create more hypotheses, $H_1, H_2, H_3, \ldots$ You update your beliefs when a new hypothesis explains your data much better than your old hypothesis:

$$\frac{P\left(D \mid H_2, X\right)}{P\left(D \mid H_1, X\right)} = \text{large number}$$

Finally, you should be far more concerned with data changing your beliefs than with ensuring data supports your beliefs, $P(H|D)$.

With these foundations set up, you're ready to start adding numbers into the mix. In the rest of Part I, you'll model your beliefs mathematically to precisely determine how and when you should change them.

Exercises

Try answering the following questions to see how well you understand Bayesian reasoning. The solutions can be found at *https://nostarch.com/ learnbayes/*.

1. Rewrite the following statements as equations using the mathematical notation you learned in this chapter:
   - The probability of rain is low
   - The probability of rain given that it is cloudy is high
   - The probability of you having an umbrella given it is raining is much greater than the probability of you having an umbrella in general.

2. Organize the data you observe in the following scenario into a mathematical notation, using the techniques we've covered in this chapter. Then come up with a hypothesis to explain this data:

> You come home from work and notice that your front door is open and the side window is broken. As you walk inside, you immediately notice that your laptop is missing.

3. The following scenario adds data to the previous one. Demonstrate how this new information changes your beliefs and come up with a second hypothesis to explain the data, using the notation you've learned in this chapter.

> A neighborhood child runs up to you and apologizes profusely for accidentally throwing a rock through your window. They claim that they saw the laptop and didn't want it stolen so they opened the front door to grab it, and your laptop is safe at their house.

2

MEASURING UNCERTAINTY

In Chapter 1 we looked at some basic reasoning tools we use intuitively to understand how data informs our beliefs. We left a crucial issue unresolved: how can we quantify these tools? In probability theory, rather than describing beliefs with terms like *very low* and *high*, we need to assign real numbers to these beliefs. This allows us to create quantitative models of our understanding of the world. With these models, we can see just how much the evidence changes our beliefs, decide when we should change our thinking, and gain a solid understanding of our current state of knowledge. In this chapter, we will apply this concept to quantify the probability of an event.

What Is a Probability?

The idea of probability is deeply ingrained in our everyday language. Whenever you say something such as "That seems unlikely!" or "I would be surprised if that's not the case" or "I'm not sure about that," you're making a claim about probability. Probability is a measurement of how strongly we believe things about the world.

In the previous chapter we used abstract, qualitative terms to describe our beliefs. To really analyze how we develop and change beliefs, we need to define exactly what a probability is by more formally quantifying $P(X)$—that is, how strongly we believe in X.

We can consider probability an extension of logic. In basic logic we have two values, true and false, which correspond to absolute beliefs. When we say something is true, it means that we are completely certain it is the case. While logic is useful for many problems, very rarely do we believe anything to be absolutely true or absolutely false; there is almost always some level of uncertainty in every decision we make. Probability allows us to extend logic to work with uncertain values between true and false.

Computers commonly represent true as 1 and false as 0, and we can use this model with probability as well. $P(X) = 0$ is the same as saying that $X =$ false, and $P(X) = 1$ is the same as $X =$ true. Between 0 and 1 we have an infinite range of possible values. A value closer to 0 means we are more certain that something is false, and a value closer to 1 means we're more certain something is true. It's worth noting that a value of 0.5 means that we are completely unsure whether something is true or false.

Another important part of logic is *negation*. When we say "not true" we mean false. Likewise, saying "not false" means true. We want probability to work the same way, so we make sure that the probability of X and the negation of the probability of X sum to 1 (in other words, values are either X, or not X). We can express this using the following equation:

$$P(X) + \neg P(X) = 1$$

NOTE   *The ¬ symbol means "negation" or "not."*

Using this logic, we can always find the negation of $P(X)$ by subtracting it from 1. So, for example, if $P(X) = 1$, then its negation, $1 - P(X)$, must equal 0, conforming to our basic logic rules. And if $P(X) = 0$, then its negation $1 - P(X) = 1$.

The next question is how to quantify that uncertainty. We could arbitrarily pick values: say 0.95 means very certain, and 0.05 means very uncertain. However, this doesn't help us determine probability much more than the abstract terms we've used before. Instead, we need to use formal methods to calculate our probabilities.

Calculating Probabilities by Counting Outcomes of Events

The most common way to calculate probability is to count outcomes of events. We have two sets of outcomes that are important. The first is all possible outcomes of an event. For a coin toss, this would be "heads" or "tails." The second is the count of the outcomes you're interested in. If you've decided that heads means you win, the outcomes you care about are those involving heads (in the case of a single coin toss, just one event). The events you're interested in can be anything: flipping a coin and getting heads, catching the flu, or a UFO landing outside your bedroom. Given these two sets of outcomes—ones you're interested in and ones you're not interested in—all we care about is the ratio of outcomes we're interested in to the total number of possible outcomes.

We'll use the simple example of a coin flip, where the only possible outcomes are the coin landing on heads or landing on tails. The first step is to make a count of all the possible events, which in this case is only two: heads or tails. In probability theory, we use Ω (the capital Greek letter omega) to indicate the set of all events:

$$\Omega = \{\text{heads, tails}\}$$

We want to know the probability of getting a heads in a single coin toss, written as $P(\text{heads})$. We therefore look at the number of outcomes we care about, 1, and divide that by the total number of possible outcomes, 2:

$$\frac{\{\text{heads}\}}{\{\text{heads, tails}\}}$$

For a single coin toss, we can see that there is one outcome we care about out of two possible outcomes. So the probability of heads is just:

$$P(\text{heads}) = \frac{1}{2}$$

Now let's ask a trickier question: what is the probability of getting at least one heads when we toss two coins? Our list of possible events is more complicated; it's not just {heads, tails} but rather all possible pairs of heads and tails:

$$\Omega = \{(\text{heads, heads}),(\text{heads, tails}),(\text{tails, tails}),(\text{tails, heads})\}$$

To figure out the probability of getting at least one heads, we look at how many of our pairs match our condition, which in this case is:

$$\{(\text{heads, heads}),(\text{heads, tails}),(\text{tails, heads})\}$$

As you can see, the set of events we care about has 3 elements, and there are 4 possible pairs we could get. This means that P(at least one heads) = 3/4.

These are simple examples, but if you can count the events you care about and the total possible events, you can come up with a quick and easy probability. As you can imagine, as examples get more complicated, manually counting each possible outcome becomes unfeasible. Solving harder probability problems of this nature often involves a field of mathematics called *combinatorics*. In Chapter 4 we'll see how we can use combinatorics to solve a slightly more complex problem.

Calculating Probabilities as Ratios of Beliefs

Counting events is useful for physical objects, but it's not so great for the vast majority of real-life probability questions we might have, such as:

- "What's the probability it will rain tomorrow?"
- "Do you think she's the president of the company?"
- "Is that a UFO!?"

Nearly every day you make countless decisions based on probability, but if someone asked you to solve "How likely do think you are to make your train on time?" you couldn't calculate it with the method just described.

This means we need another approach to probability that can be used to reason about these more abstract problems. As an example, suppose you're chatting about random topics with a friend. Your friend asks if you've heard of the Mandela effect and, since you haven't, proceeds to tell you: "It's this weird thing where large groups of people misremember events. For example, many people recall Nelson Mandela dying in prison in the 80s. But the wild thing is that he was released from prison, became president of South Africa, and didn't die until 2013!" Skeptically, you turn to your friend and say, "That sounds like internet pop psychology. I don't think anyone seriously misremembered that; I bet there's not even a Wikipedia entry on it!"

From this, you want to measure P(No Wikipedia article on Mandela effect). Let's assume you are in an area with no cell phone reception, so you can't quickly verify the answer. You have a high certainty of your belief that there is no such article, and therefore you want to assign a high probability for this belief, but you need to formalize that probability by assigning it a number from 0 to 1. Where do you start?

You decide to put your money where your mouth is, telling your friend: "There's no way that's real. How about this: *you give me $5 if there is no article on the Mandela effect, and I'll give you $100 if there is one!*" Making bets is a practical way that we can express how strongly we hold our beliefs. You believe that the article's existence is so unlikely that you'll give your friend $100 if you are wrong and only get $5 from them if you are right. Because we're talking about quantitative values regarding our beliefs, we can start to figure out an exact probability for your belief that there is no Wikipedia article on the Mandela effect.

Using Odds to Determine Probability

Your friend's hypothesis is that there is an article about the Mandela effect: $H_{article}$. And you have an alternate hypothesis: $H_{no\ article}$.

We don't have concrete probabilities yet, but your bet expresses how strongly you believe in your hypothesis by giving the *odds* of the bet. Odds are a common way to represent beliefs as a ratio of how much you would be willing to pay if you were wrong about the outcome of an event to how much you'd want to receive for being correct. For example, say the odds of a horse winning a race are 12 to 1. That means if you pay $1 to take the bet, the track will pay you $12 if the horse wins. While odds are commonly expressed as "*m* to *n*" we can also view them as a simple ratio: *m/n*. There is a direct relationship between odds and probabilities.

We can express your bet in terms of odds as "100 to 5." So how can we turn this into probability? Your odds represent how many times more strongly you believe there *isn't* an article than you believe there *is* an article. We can write this as the ratio of your belief in there being no article, $P(H_{no\ article})$, to your friend's belief that there is one, $P(H_{article})$, like so:

$$\frac{P(H_{no\ article})}{P(H_{article})} = \frac{100}{5} = 20$$

From the ratio of these two hypotheses, we can see that your belief in the hypothesis that there is no article is 20 times greater than your belief in your friend's hypothesis. We can use this fact to work out the exact probability for your hypothesis using some high school algebra.

Solving for the Probabilities

We start writing our equation in terms of the probability of your hypothesis, since this is what we are interested in knowing:

$$P(H_{no\ article}) = 20 \times P(H_{article})$$

We can read this equation as "The probability that there is no article is 20 times greater than the probability there is an article."

There are only two possibilities: either there is a Wikipedia article on the Mandela effect or there isn't. Because our two hypotheses cover all possibilities, we know that the probability of an *article* is just 1 minus the probability of *no article*, so we can substitute $P(H_{article})$ with its value in terms of $P(H_{no\ article})$ in our equation like so:

$$P(H_{no\ article}) = 20 \times (1 - P(H_{no\ article}))$$

Next we can expand $20 \times (1 - P(H_{\text{no article}}))$ by multiplying both parts in the parentheses by 20 and we get:

$$P(H_{\text{no article}}) = 20 - 20 \times P(H_{\text{no article}})$$

We can remove the $P(H_{\text{no article}})$ term from the right side of the equation by adding $20 \times P(H_{\text{no article}})$ to both sides to isolate $P(H_{\text{no article}})$ on the left side of the equation:

$$21 \times P(H_{\text{no article}}) = 20$$

And we can divide both sides by 21, finally arriving at:

$$P(H_{\text{no article}}) = \frac{20}{21}$$

Now you have a nice, clearly defined value between 0 and 1 to assign as a concrete, quantitative probability to your belief in the hypothesis that there is no article on the Mandela effect. We can generalize this process of converting odds to probability using the following equation:

$$P(H) = \frac{O(H)}{1 + O(H)}$$

Often in practice, when you're confronted with assigning a probability to an abstract belief, it can be very helpful to think of how much you would bet on that belief. You would likely take a billion to 1 bet that the sun will rise tomorrow, but you might take much lower odds for your favorite baseball team winning. In either case, you can calculate an exact number for the probability of that belief using the steps we just went through.

Measuring Beliefs in a Coin Toss

We now have a method for determining the probability of abstract ideas using odds, but the real test of the robustness of this method is whether or not it still works with our coin toss, which we calculated by counting outcomes. Rather than thinking about a coin toss as an *event*, we can rephrase the question as "How strongly do I believe the next coin toss will be heads?" Now we're not talking about $P(\text{heads})$ but rather a hypothesis or belief about the coin toss, $P(H_{\text{heads}})$.

Just like before, we need an alternate hypothesis to compare our belief with. We could say the alternate hypothesis is simply not getting heads $H_{\neg\text{heads}}$, but the option of getting tails H_{tails} is closer to our everyday language, so we'll use that. At the end of the day what we care about most is making sense. However, it is important for this discussion to acknowledge that:

$$H_{\text{tails}} = H_{\neg\text{heads}}, \text{ and } P(H_{\text{tails}}) = 1 - P(H_{\text{heads}})$$

We can look at how to model our beliefs as the ratio between these competing hypotheses:

$$\frac{P(H_{\text{heads}})}{P(H_{\text{tails}})} = ?$$

Remember that we want to read this as "How many times greater do I believe that the outcome will be heads than I do that it will be tails?" As far as bets go, since each outcome is equally uncertain, the only fair odds are 1 to 1. Of course, we can pick any odds as long as the two values are equal: 2 to 2, 5 to 5, or 10 to 10. All of these have the same ratio:

$$\frac{P(H_{\text{heads}})}{P(H_{\text{tails}})} = \frac{10}{10} = \frac{5}{5} = \frac{2}{2} = \frac{1}{1} = 1$$

Given that the ratio of these is always the same, we can simply repeat the process we used to calculate the probability of there being no Wikipedia article on the Mandela effect. We know that our probability of heads and probability of tails must sum to 1, and we know that the ratio of these two probabilities is also 1. So, we have two equations that describe our probabilities:

$$P(H_{\text{heads}}) + P(H_{\text{tails}}) = 1, \text{ and } \frac{P(H_{\text{heads}})}{P(H_{\text{tails}})} = 1$$

If you walk through the process we used when reasoning about the Mandela effect, solving in terms of $P(H_{\text{heads}})$ you should find the only possible solution to this problem is 1/2. This is exactly the same result we arrived at with our first approach to calculating probabilities of events, and it proves that our method for calculating the probability of a belief is robust enough to use for the probability of events!

With these two methods in hand, it's reasonable to ask which one you should use in which situation. The good news is, since we can see they are equivalent, you can use whichever method is easiest for a given problem.

Wrapping Up

In this chapter we explored two different types of probabilities: those of events and those of beliefs. We define probability as the ratio of the outcome(s) we care about to the number of all possible outcomes.

While this is the most common definition of probability, it is difficult to apply to beliefs because most practical, everyday probability problems do not have clear-cut outcomes and so aren't intuitively assigned discrete numbers.

To calculate the probability of beliefs, then, we need to establish how many times more we believe in one hypothesis over another. One good test of this is how much you would be willing to bet on your belief—for example, if you made a bet with a friend in which you'd give them $1,000 for proof that UFOs exist and would receive only $1 from them for proof that UFOs don't exist. Here you are saying you believe UFOs do not exist 1,000 times more than you believe they do exist.

With these tools in hand, you can calculate the probability for a wide range of problems. In the next chapter you'll learn how you can apply the basic operators of logic, AND and OR, to our probabilities. But before moving on, try using what you've learned in this chapter to complete the following exercises.

Exercises

Try answering the following questions to make sure you understand how we can assign real values between 0 and 1 to our beliefs. Solutions to the questions can be found at *https://nostarch.com/learnbayes/*.

1. What is the probability of rolling two six-sided dice and getting a value greater than 7?

2. What is the probability of rolling three six-sided dice and getting a value greater than 7?

3. The Yankees are playing the Red Sox. You're a diehard Sox fan and bet your friend they'll win the game. You'll pay your friend $30 if the Sox lose and your friend will have to pay you only $5 if the Sox win. What is the probability you have intuitively assigned to the belief that the Red Sox will win?

3

THE LOGIC OF UNCERTAINTY

In Chapter 2, we discussed how probabilities are an extension of the true and false values in logic and are expressed as values between 1 and 0. The power of probability is in the ability to express an infinite range of possible values between these extremes. In this chapter, we'll discuss how the rules of logic, based on these logical operators, also apply to probability. In traditional logic, there are three important operators:

- AND
- OR
- NOT

With these three simple operators we can reason about any argument in traditional logic. For example, consider this statement: *If it is raining AND I am going outside, I will need an umbrella.* This statement contains just one logical operator: AND. Because of this operator we know that if it's true that it is raining, AND it is true that I am going outside, I'll need an umbrella.

We can also phrase this statement in terms of our other operators: *If it is NOT raining OR if I am NOT going outside, I will NOT need an umbrella.* In this case we are using basic logical operators and facts to make a decision about when we do and don't need an umbrella.

However, this type of logical reasoning works well only when our facts have absolute true or false values. This case is about deciding whether I need an umbrella *right now*, so we can know for certain if it's currently raining and whether I'm going out, and therefore I can easily determine if I need an umbrella. Suppose instead we ask, "Will I need an umbrella tomorrow?" In this case our facts become uncertain, because the weather forecast gives me only a probability for rain tomorrow and I may be uncertain whether or not I need to go out.

This chapter will explain how we can extend our three logical operators to work with probability, allowing us to reason about uncertain information the same way we can with facts in traditional logic. We've already seen how we can define NOT for probabilistic reasoning:

$$\neg P(X) = 1 - P(X)$$

In the rest of this chapter we'll see how we can use the two remaining operators, AND and OR, to combine probabilities and give us more accurate and useful data.

Combining Probabilities with AND

In statistics we use AND to talk about the probability of combined events. For example, the probability of:

- Rolling a 6 AND flipping a heads
- It raining AND you forgetting your umbrella
- Winning the lottery AND getting struck by lightning

To understand how we can define AND for probability, we'll start with a simple example involving a coin and a six-sided die.

Solving a Combination of Two Probabilities

Suppose we want to know the probability of getting a heads in a coin flip AND rolling a 6 on a die. We know that the probability of *each* of these events individually is:

$$P(\text{heads}) = \frac{1}{2}, \ P(\text{six}) = \frac{1}{6}$$

Now we want to know the probability of *both* of these things occurring, written as:

$$P(\text{heads, six}) = ?$$

We can calculate this the same way we did in Chapter 2: we count the outcomes we care about and divide that by the total outcomes.

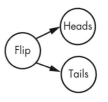

For this example, let's imagine these events happening in sequence. When we flip the coin we have two possible outcomes, heads and tails, as depicted in Figure 3-1.

Figure 3-1: Visualizing the two possible outcomes from a coin toss as distinct paths

Now, for each possible coin flip there are six possible results for the roll of our die, as depicted in Figure 3-2.

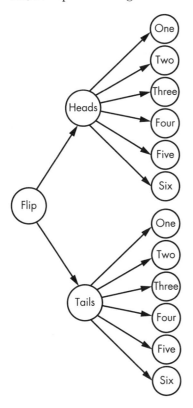

Figure 3-2: Visualizing the possible outcomes from a coin toss and the roll of a die

Using this visualization, we can just count our possible solutions. There are 12 possible outcomes of flipping a coin and rolling a die, and we care about only one of these outcomes, so:

$$P(\text{heads, six}) = \frac{1}{12}$$

Now we have a solution for this particular problem. However, what we really want is a general rule that will help us calculate this for any number of probability combinations. Let's see how to expand our solution.

Applying the Product Rule of Probability

We'll use the same problem for this example: what is the probability of flipping a heads and rolling a 6? First we need to figure out the probability of flipping a heads. Looking at our branching paths, we can figure out how many paths split off given the probabilities. We care only about the paths that include heads. Because the probability of heads is 1/2, we eliminate half of our possibilities. Then, if we look only at the remaining branch of possibilities for the heads, we can see that there is only a 1/6 chance of getting the result we want: rolling a 6 on a six-sided die. In Figure 3-3 we can visualize this reasoning and see that there is only one outcome we care about.

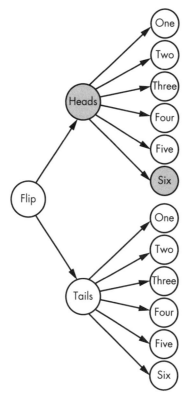

Figure 3-3: Visualizing the probability of both getting a heads and rolling a 6

If we multiply these two probabilities, we can see that:

$$\frac{1}{2} \times \frac{1}{6} = \frac{1}{12}$$

This is exactly the answer we had before, but rather than counting all possible events, we counted only the probabilities of the events we care about by following along the branches. This is easy enough to do visually for such a simple problem, but the real value of showing you this is that it illustrates a general rule for combining probabilities with AND:

$$P(A,B) = P(A) \times P(B)$$

Because we are multiplying our results, also called taking the *product* of these results, we refer to this as the *product rule* of probability.

This rule can then be expanded to include more probabilities. If we think of $P(A,B)$ as a single probability, we can combine it with a third probability, $P(C)$, by repeating this process:

$$P(P(A,B),C) = P(A,B) \times P(C) = P(A) \times P(B) \times P(C)$$

So we can use our product rule to combine an unlimited number of events to get our final probability.

Example: Calculating the Probability of Being Late

Let's look at an example of using the product rule for a slightly more complex problem than rolling dice or flipping coins. Suppose you promised to meet a friend for coffee at 4:30 on the other side of town, and you plan to take public transportation. It's currently 3:30. Thankfully the station you're at has both a train and bus that can take you where you need to go:

- The next bus comes at 3:45 and takes 45 minutes to get you to the coffee shop.
- The next train comes at 3:50, and will get you within a 10-minute walk in 30 minutes.

Both the train and the bus will get you there at 4:30 exactly. Because you're cutting it so close, any delay will make you late. The good news is that, since the bus arrives before the train, if the bus is late and the train is not you'll still be on time. If the bus is on time and the train is late, you'll also be fine. The only situation that will make you late is if both the bus and the train are late to arrive. How can you figure out the probability of being late?

First, you need to establish the probability of both the train being late and the bus being late. Let's assume the local transit authority publishes these numbers (later in the book, you'll learn how to estimate this from data).

$$P\left(\text{Late}_{\text{train}}\right) = 0.15$$
$$P\left(\text{Late}_{\text{bus}}\right) = 0.2$$

The published data tells us that 15 percent of the time the train is late, and 20 percent of the time the bus is late. Since you'll be late only if *both* the bus and the train are late, we can use the product rule to solve this problem:

$$P\left(\text{Late}\right) = P\left(\text{Late}_{\text{train}}\right) \times P\left(\text{Late}_{\text{bus}}\right) = 0.15 \times 0.2 = 0.03$$

Even though there's a pretty reasonable chance that either the bus or the train will be late, the probability that they will both be late is significantly less, at only 0.03. We can also say there is a 3 percent chance that both will be late. With this calculation done, you can be a little less stressed about being late.

Combining Probabilities with OR

The other essential rule of logic is combining probabilities with OR, some examples of which include:

- Catching the flu OR getting a cold
- Flipping a heads on a coin OR rolling a 6 on a die
- Getting a flat tire OR running out of gas

The probability of one event OR another event occurring is slightly more complicated because the events can either be mutually exclusive or not mutually exclusive. Events are *mutually exclusive* if one event happening implies the other possible events cannot happen. For example, the possible outcomes of rolling a die are mutually exclusive because a single roll cannot yield both a 1 and a 6. However, say a baseball game will be cancelled if it is either raining or the coach is sick; these events are *not* mutually exclusive because it is perfectly possible that the coach is sick and it rains.

Calculating OR for Mutually Exclusive Events

The process of combining two events with OR feels logically intuitive. If you're asked, "What is the probability of getting heads or tails on a coin toss?" you would say, "1." We know that:

$$P\left(\text{heads}\right) = \frac{1}{2}, \; P\left(\text{tails}\right) = \frac{1}{2}$$

Intuitively, we might just add the probability of these events together. We know this works because heads and tails are the only possible outcomes, and the probability of all possible outcomes must equal 1. If the probabilities of all possible events did not equal 1, then we would have some outcome that was missing. So how do we know that there would need to be a missing outcome if the sum was less than 1?

Suppose we know that the probability of heads is $P(\text{heads}) = 1/2$, but someone claimed that the probability of tails was $P(\text{tails}) = 1/3$. We also know from before that the probability of not getting heads must be:

$$\text{NOT } P(\text{heads}) = 1 - \frac{1}{2} = \frac{1}{2}$$

Since the probability of not getting heads is 1/2 and the claimed probability for tails is only 1/3, either there is a missing event or our probability for tails is incorrect.

From this we can see that, as long as events are mutually exclusive, we can simply add up all of the probabilities of each possible event to get the probability of either event happening to calculate the probability of one event OR the other. Another example of this is rolling a die. We know that the probability of rolling a 1 is 1/6, and the same is true for rolling a 2:

$$P(\text{one}) = \frac{1}{6}, \, P(\text{two}) = \frac{1}{6}$$

So we can perform the same operation, adding the two probabilities, and see that the combined probability of rolling either a 1 OR a 2 is 2/6, or 1/3:

$$P(\text{one}) + P(\text{two}) = \frac{2}{6} = \frac{1}{3}$$

Again, this makes intuitive sense.

This addition rule applies only to combinations of *mutually exclusive* outcomes. In probabilistic terms, mutually exclusive means that:

$$P(A) \text{ AND } P(B) = 0$$

That is, the probability of getting both A AND B together is 0. We see that this holds for our examples:

- It is impossible to flip one coin and get both heads and tails.
- It is impossible to roll both a 1 and a 2 on a single roll of a die.

To really understand combining probabilities with OR, we need to look at the case where events are *not* mutually exclusive.

Using the Sum Rule for Non–Mutually Exclusive Events

Again using the example of rolling a die and flipping a coin, let's look at the probability of either flipping heads OR rolling a 6. Many newcomers to probability may naively assume that adding probabilities will work in this case as well. Given that we know that $P(\text{heads}) = 1/2$ and $P(\text{six}) = 1/6$, it might initially seem plausible that the probability of either of these events is simply 4/6. It becomes obvious that this doesn't work, however, when we consider the possibility of either flipping a heads or rolling a number less than 6. Because $P(\text{less than six}) = 5/6$, adding these probabilities together gives us 8/6, which is greater than 1! Since this violates the rule that probabilities must be between 0 and 1, we must have made a mistake.

The trouble is that flipping a heads and rolling a 6 are not mutually exclusive. As we know from earlier in the chapter, $P(\text{heads, six}) = 1/12$. Because the probability of both events happening at the same time is not 0, we know they are, by definition, not mutually exclusive.

The reason that adding our probabilities doesn't work for non–mutually exclusive events is that doing so doubles the counting of events where both things happen. As an example of overcounting, let's look at all of the outcomes of our combined coin toss and die roll that contain heads:

Heads — 1

Heads — 2

Heads — 3

Heads — 4

Heads — 5

Heads — 6

These outcomes represent 6 out of the 12 possible outcomes, which we expect since $P(\text{heads}) = 1/2$. Now let's look at all outcomes that include rolling a 6:

Heads — 6

Tails — 6

These outcomes represent the 2 out of 12 possible outcomes that will result in us rolling a 6, which again we expect because $P(\text{six}) = 1/6$. Since there are six outcomes that satisfy the condition of flipping a heads and two that satisfy the condition of rolling a 6, we might be tempted to say that there are eight outcomes that represent getting either heads or rolling a 6. However, we would be double-counting because *Heads — 6* appears in both lists. There are, in fact, only 7 out of 12 unique outcomes. If we naively add $P(\text{heads})$ and $P(\text{six})$, we end up overcounting.

To correct our probabilities, we must add up all of our probabilities and then subtract the probability of both events occurring. This leads us to the rule for combining non–mutually exclusive probabilities with OR, known as the *sum rule* of probability:

$$P(A) \text{ OR } P(B) = P(A) + P(B) - P(A,B)$$

We add the probability of each event happening and then subtract the probability of both events happening, to ensure we are not counting these probabilities twice since they are a part of both $P(A)$ and $P(B)$. So, using our die roll and coin toss example, the probability of rolling a number less than 6 or flipping a heads is:

$$P(\text{heads}) \text{ OR } P(\text{six}) = P(\text{heads}) + P(\text{six}) - P(\text{heads, six}) = \frac{1}{2} + \frac{1}{6} - \frac{1}{12} = \frac{7}{12}$$

Let's take a look at a final OR example to really cement this idea.

Example: Calculating the Probability of Getting a Hefty Fine

Imagine a new scenario. You were just pulled over for speeding while on a road trip. You realize you haven't been pulled over in a while and may have forgotten to put either your new registration or your new insurance card in the glove box. If either one of these is missing, you'll get a more expensive ticket. Before you open the glove box, how can you assign a probability that you'll have forgotten one or the other of your cards and you'll get the higher ticket?

You're pretty confident that you put your registration in the car, so you assign a 0.7 probability to your registration being in the car. However, you're also pretty sure that you left your insurance card on the counter at home, so you assign only a 0.2 chance that your new insurance card is in the car. So we know that:

$$P(\text{registration}) = 0.7$$
$$P(\text{insurance}) = 0.2$$

However, these values are the probabilities that you *do* have these things in the glove box. You're worried about whether either one is *missing*. To get the probabilities of missing items, we simply use negation:

$$P(\text{Missing}_{\text{reg}}) = 1 - P(\text{registration}) = 0.3$$
$$P(\text{Missing}_{\text{ins}}) = 1 - P(\text{insurance}) = 0.8$$

If we try using our addition method, instead of the complete sum rule, to get the combined probability, we see that we have a probability greater than 1:

$$P(\text{Missing}_{\text{reg}}) + P(\text{Missing}_{\text{ins}}) = 1.1$$

This is because these events are non–mutually exclusive: it's entirely possible that you have forgotten both cards. Therefore, using this method we're double-counting. That means we need to figure out the probability that you're missing both cards so we can subtract it. We can do this with the product rule:

$$P\left(\text{Missing}_{\text{reg}}, \text{Missing}_{\text{ins}}\right) = 0.24$$

Now we can use the sum rule to determine the probability that either one of these cards is missing, just as we worked out the probability of a flipping a heads or rolling a 6:

$$P\left(\text{Missing}\right) = P\left(\text{Missing}_{\text{reg}}\right) + P\left(\text{Missing}_{\text{ins}}\right) - P\left(\text{Missing}_{\text{reg}}, \text{Missing}_{\text{ins}}\right) = 0.86$$

With an 0.86 probability that one of these important pieces of paper is missing from your glove box, you should make sure to be extra nice when you greet the officer!

Wrapping Up

In this chapter you developed a complete logic of uncertainty by adding rules for combining probabilities with AND and OR. Let's review the logical rules we have covered so far.

In Chapter 2, you learned that probabilities are measured on a scale of 0 to 1, 0 being *false* (definitely not going to happen), and 1 being *true* (definitely going to happen). The next important logical rule involves combining two probabilities with AND. We do this using the product rule, which simply states that to get the probability of two events occurring together, $P(A)$ and $P(B)$, we just multiply them together:

$$P\left(A, B\right) = P\left(A\right) \times P\left(B\right)$$

The final rule involves combining probabilities with OR using the sum rule. The tricky part of the sum rule is that if we add non–mutually exclusive probabilities, we'll end up overcounting for the case where they both occur, so we have to subtract the probability of both events occurring together. The sum rule uses the product rule to solve this (remember, for mutually exclusive events, $P(A, B) = 0$):

$$P\left(A \text{ OR } B\right) = P\left(A\right) + P\left(B\right) - P\left(A, B\right)$$

These rules, along with those covered in Chapter 2, allow us to express a very large range of problems. We'll be using these as the foundation for our probabilistic reasoning throughout the rest of the book.

Exercises

Try answering the following questions to make sure you understand the rules of logic as they apply to probability. The solutions can be found at *https://nostarch.com/learnbayes/*.

1. What is the probability of rolling a 20 three times in a row on a 20-sided die?

2. The weather report says there's a 10 percent chance of rain tomorrow, and you forget your umbrella half the time you go out. What is the probability that you'll be caught in the rain without an umbrella tomorrow?

3. Raw eggs have a 1/20,000 probability of having salmonella. If you eat two raw eggs, what is the probability you ate a raw egg with salmonella?

4. What is the probability of either flipping two heads in two coin tosses or rolling three 6s in three six-sided dice rolls?

4

CREATING A BINOMIAL
PROBABILITY DISTRIBUTION

In Chapter 3, you learned some basic rules of probability corresponding to the common logical operators: AND, OR, and NOT. In this chapter we're going to use these rules to build our first *probability distribution*, a way of describing all possible events and the probability of each one happening. Probability distributions are often visualized to make statistics more palatable to a wider audience. We'll arrive at our probability distribution by defining a function that *generalizes* a particular group of probability problems, meaning we'll create a distribution to calculate the probabilities for a whole range of situations, not just one particular case.

We generalize in this way by looking at the common elements of each problem and abstracting them out. Statisticians use this approach to make solving a wide range of problems much easier. This can be especially useful when problems are very complex, or some of the necessary details may be unknown. In these cases, we can use well-understood probability distributions as estimates for real-world behavior that we don't fully understand.

Probability distributions are also very useful for asking questions about ranges of possible values. For example, we might use a probability distribution to determine the probability that a customer makes between \$30,000 and \$45,000 a year, the probability of an adult being taller than 6' 10", or the probability that between 25 percent and 35 percent of people who visit a web page will sign up for an account there. Many probability distributions involve very complex equations and can take some time to get used to. However, all the equations for probability distributions are derived from the basic rules of probability covered in the previous chapters.

Structure of a Binomial Distribution

The distribution you'll learn about here is the *binomial distribution*, used to calculate the probability of a certain number of successful outcomes, given a number of trials and the probability of the successful outcome. The "bi" in the term *binomial* refers to the two possible outcomes that we're concerned with: an event happening and an event *not* happening. If there are more than two outcomes, the distribution is called *multinomial*. Example problems that follow a binomial distribution include the probability of:

- Flipping two heads in three coin tosses
- Buying 1 million lottery tickets and winning at least once
- Rolling fewer than three 20s in 10 rolls of a 20-sided die

Each of these problems shares a similar structure. Indeed, all binomial distributions involve three *parameters*:

k The number of outcomes we care about

n The total number of trials

p The probability of the event happening

These parameters are the inputs to our distribution. So, for example, when we're calculating the probability of flipping two heads in three coin tosses:

- $k = 2$, the number of events we care about, in this case flipping a heads
- $n = 3$, the number times the coin is flipped
- $p = 1/2$, the probability of flipping a heads in a coin toss

We can build out a binomial distribution to generalize this kind of problem, so we can easily solve any problem involving these three parameters. The shorthand notation to express this distribution looks like this:

$$B(k; n, p)$$

For the example of three coin tosses, we would write $B(2; 3, 1/2)$. The B is short for *binomial* distribution. Notice that the k is separated from the other parameters by a semicolon. This is because when we are talking about a distribution of values, we usually care about all values of k for a fixed n and p. So $B(k; n, p)$ denotes each value in our distribution, but the entire distribution is usually referred to by simply $B(n, p)$.

Let's take a look at this more closely and see how we can build a function that allows us to generalize all of these problems into the binomial distribution.

Understanding and Abstracting Out the Details of Our Problem

One of the best ways to see how creating distributions can simplify your probabilities is to start with a concrete example and try to solve that, and then abstract out as many of the variables as you can. We'll continue with the example of calculating the probability of flipping two heads in three coin tosses.

Since the number of possible outcomes is small, we can quickly figure out the results we care about with just pencil and paper. There are three possible outcomes with two heads in three tosses:

HHT, HTH, THH

Now it may be tempting to just solve this problem by enumerating all the other possible outcomes and dividing the number we care about by the total number of possible outcomes (in this case, 8). That would work fine for solving *just* this problem, but our aim here is to solve any problem that involves desiring a set of outcomes, from a number of trials, with a given probability that the event occurs. If we did not generalize and solved only this one instance of the problem, changing these parameters would mean we have to solve the new problem again. For example, just saying, "What is the probability of getting two heads in *four* coin tosses?" means we need to come up with yet another unique solution. Instead, we'll use the rules of probability to reason about this problem.

To start generalizing, we'll break this problem down into smaller pieces we can solve right now, and reduce those pieces into manageable equations. As we build up the equations, we'll put them together to create a generalized function for the binomial distribution.

The first thing to note is that each outcome we care about will have the *same* probability. Each outcome is just a *permutation*, or reordering, of the others:

$$P\big(\{\text{heads, heads, tails}\}\big) = P\big(\{\text{heads, tails, heads}\}\big) = P\big(\{\text{tails, heads, heads}\}\big)$$

Since this is true, we'll simply call it:

$$P(\text{Desired Outcome})$$

There are three outcomes, but only one of them can possibly happen and we don't care which. And because it's only possible for one outcome to occur, we know that these are mutually exclusive, denoted as:

$$P(\{\text{heads, heads, tails}\}, \{\text{heads, tails, heads}\}, \{\text{tails, heads, heads}\}) = 0$$

This makes using the sum rule of probability easy. Now we can summarize this nicely as:

$$P(\{\text{heads, heads, tails}\} \text{ or } \{\text{heads, tails, heads}\} \text{ or } \{\text{tails, heads, heads}\}) =$$
$$P(\text{Desired Outcome}) + P(\text{Desired Outcome}) + P(\text{Desired Outcome})$$

Of course adding these three is just the same as:

$$3 \times P(\text{Desired Outcome})$$

We've got a condensed way of referencing the outcomes we care about, but the trouble as far as generalizing goes is that the value 3 is specific to this problem. We can fix this by simply replacing 3 with a variable called N_{outcomes}. This leaves us with a pretty nice generalization:

$$B(k; n, p) = N_{\text{outcomes}} \times P(\text{Desired Outcome})$$

Now we have to figure out two subproblems: how to count the number of outcomes we care about, and how to determine the probability for a single outcome. Once we have these fleshed out, we'll be all set!

Counting Our Outcomes with the Binomial Coefficient

First we need to figure out how many outcomes there are for a given k (the outcomes we care about) and n (the number of trials). For small numbers we can simply count. If we were looking at four heads in five coin tosses, we know there are five outcomes we care about:

$$\text{HHHHT, HTHHH, HHTHH, HHHTH, HHHHT}$$

But it doesn't take much for this to become too difficult to do by hand—for example, "What is the probability of rolling two 6s in three rolls of a six-sided die?"

This is still a binomial problem, because the only two possible outcomes are getting a 6 or not getting a 6, but there are far more events that count as "not getting a 6." If we start enumerating we quickly see this gets tedious, even for a small problem involving just three rolls of a die:

$$6 - 6 - 1$$
$$6 - 6 - 2$$
$$6 - 6 - 3$$
$$. . .$$
$$4 - 6 - 6$$
$$. . .$$
$$5 - 6 - 6$$
$$. . .$$

Clearly, enumerating all of the possible solutions will not scale to even reasonably trivial problems. The solution is combinatorics.

Combinatorics: Advanced Counting with the Binomial Coefficient

We can gain some insight into this problem if we take a look at a field of mathematics called *combinatorics*. This is simply the name for a kind of advanced counting.

There is a special operation in combinatorics, called the *binomial coefficient*, that represents counting the number of ways we can select k from n—that is, selecting the outcomes we care about from the total number of trials. The notation for the binomial coefficient looks like this:

$$\binom{n}{k}$$

We read this expression as "n choose k." So, for our example, we would represent "in three tosses choose two heads" as:

$$\binom{3}{2}$$

The definition of this operation is:

$$\binom{n}{k} = \frac{n!}{k! \times (n-k)!}$$

The ! means *factorial*, which is the product of all the numbers up to and including the number before the ! symbol, so $5! = (5 \times 4 \times 3 \times 2 \times 1)$.

Most mathematical programming languages indicate the binomial coefficient using the choose() function. For example, with the mathematical language R, we would compute the binomial coefficient for the case of flipping two heads in three tosses with the following call:

```
choose(3,2)
>>3
```

With this general operation for calculating the number of outcomes we care about, we can update our generalized formula like so:

$$B(k;n,p) = \binom{n}{k} \times P(\text{Desired Outcome})$$

Recall that P(Desired Outcome) is the probability of any one of the combinations of getting two heads in three coin tosses. In the preceding equation, we use this value as a placeholder, but we don't actually know how to calculate what this value is. The only missing piece of our puzzle is solving P(Single Outcome). After that, we'll be able to easily generalize an entire class of problems!

Calculating the Probability of the Desired Outcome

All we have left to figure out is the P(Desired Outcome), which is the probability of any of the possible events we care about. So far we've been using P(Desired Outcome) as a variable to help organize our solution to this problem, but now we need to figure out exactly how to calculate this value. Let's look at the probability of getting two heads in five tosses. We'll focus on a single case of an outcome that meets this condition: HHTTT.

We know the probability of flipping a heads in a single toss is $1/2$, but to generalize the problem we'll work with it as P(heads) so we won't be stuck with a fixed value for our probability. Using the product rule and negation from the previous chapter, we can describe this problem as:

$$P(\text{heads, heads, not heads, not heads, not heads})$$

Or, more verbosely, as: "The probability of flipping heads, heads, not heads, not heads, and not heads."

Negation tells us that we can represent "not heads" as $1 - P$(heads). Then we can use the product rule to solve the rest:

$$P(\text{heads, heads, not heads, not heads, not heads}) =$$
$$P(\text{heads}) \times P(\text{heads}) \times (1 - P(\text{heads})) \times (1 - P(\text{heads})) \times (1 - P(\text{heads}))$$

Let's simplify the multiplication by using exponents:

$$P(\text{heads})^2 \times (1 - P(\text{heads}))^3$$

If we put this all together, we see that:

$$(\text{two heads in five tosses}) = P(\text{heads})^2 \times (1 - P(\text{heads}))^3$$

You can see that the exponents for $P(\text{heads})^2$ and $1 - P(\text{heads})^3$ are just the number of heads and the number of not heads in that scenario. These equate to k, the number of outcomes we care about, and $n - k$, the number of trials minus the outcomes we care about. We can put all of this together to create this much more general formula, which eliminates numbers specific to this case:

$$\binom{n}{k} \times P(\text{heads})^k \times (1 - P(\text{heads}))^{n-k}$$

Now let's generalize it for any probability, not just heads, by replacing $P(\text{heads})$ with just p. This gives us a general solution for k, the number of outcomes we care about; n, the number of trials; and p, the probability of the individual outcome:

$$B(k; n, p) = \binom{n}{k} \times p^k \times (1 - p)^{n-k}$$

Now that we have this equation, we can solve any problem related to outcomes of a coin toss. For example, we could calculate the probability of flipping exactly 12 heads in 24 coin tosses like so:

$$B\left(12; 24, \frac{1}{2}\right) = \binom{24}{12} \times \frac{1}{2}^{12} \times \left(1 - \frac{1}{2}\right)^{24-12} = 0.1612$$

Before you learned about the binomial distribution, solving this problem would have been much trickier!

This formula, which is the basis of the binomial distribution, is called a *Probability Mass Function (PMF)*. The *mass* part of the name comes from the fact that we can use it to calculate the amount of probability for *any* given k using a fixed n and p, so this is the mass of our probability.

For example, we can plug in all the possible values for k in 10 coin tosses into our PMF and visualize what the binomial distribution looks like for all possible values, as shown in Figure 4-1.

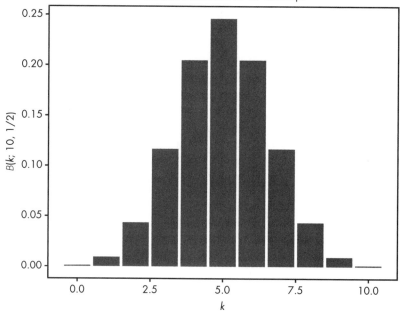

Figure 4-1: Bar graph showing the probability of getting k in 10 coin flips

We can also look at the same distribution for the probability of getting a 6 when rolling a six-sided die 10 times, shown in Figure 4-2.

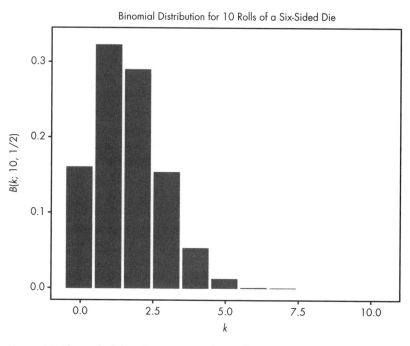

Figure 4-2: The probability of getting a 6 when rolling a six-sided die 10 times

As you can see, a probability distribution is a way of generalizing an entire class of problems. Now that we have our distribution, we have a powerful method to solve a wide range of problems. But always remember that we derived this distribution from our simple rules of probability. Let's put it to the test.

Example: Gacha Games

Gacha games are a genre of mobile games, particularly popular in Japan, in which players are able to purchase virtual cards with in-game currency. The catch is that all cards are given at random, so when players purchase cards they can't choose which ones they receive. Since not all cards are equally desirable, players are encouraged to keep pulling cards from the stack until they hit the one they want, in a fashion similar to a slot machine. We'll see how the binomial distribution can help us to decide to take a particular risk in an imaginary Gacha game.

Here's the scenario. You have a new mobile game, *Bayesian Battlers*. The current set of cards you can pull from is called a *banner*. The banner contains some average cards and some featured cards that are more valuable. As you may suspect, all of the cards in *Bayesian Battlers* are famous probabilists and statisticians. The top cards in this banner are as follows, each with its respective probability of being pulled:

- Thomas Bayes: 0.721%
- E. T. Jaynes: 0.720%
- Harold Jeffreys: 0.718%
- Andrew Gelman: 0.718%
- John Kruschke: 0.714%

These featured cards account for only 0.03591 of the total probability. Since probability must sum to 1, the chance of pulling the less desirable cards is the other 0.96409. Additionally, we treat the pile of cards that we pull from as effectively infinite, meaning that pulling a specific card does not change the probability of getting any other card—the card you pull here does not then disappear from the pile. This is different than if you were to pull a physical card from a single deck of cards without shuffling the card back in.

You really want the E. T. Jaynes card to complete your elite Bayesian team. Unfortunately, you have to purchase the in-game currency, Bayes Bucks, in order to pull cards. It costs one Bayes Buck to pull one card, but there's a special on right now allowing you to purchase 100 Bayes Bucks for only \$10. That's the maximum you are willing to spend on this game, and *only* if you have at least an even chance of pulling the card you want. This means you'll buy the Bayes Bucks only if the probability of getting that awesome E. T. Jaynes card is greater than or equal to 0.5.

Of course we can plug our probability of getting the E. T. Jaynes card into our formula for the binomial distribution to see what we get:

$$\binom{100}{1} \times 0.00720^1 \times (1 - 0.00720)^{99} = 0.352$$

Our result is less than 0.5, so we should give up. But wait—we forgot something very important! In the preceding formula we calculated only the probability of getting *exactly one* E. T. Jaynes card. But we might pull two E. T. Jaynes cards, or even three! So what we really want to know is the probability of getting one or more. We could write this out as:

$$\binom{100}{1} \times 0.00720^1 \times (1 - 0.00720)^{99} + \binom{100}{2} \times 0.00720^2 \times (1 - 0.00720)^{98} +$$
$$\binom{100}{3} \times 0.00720^3 \times (1 - 0.00720)^{97} \dots$$

And so on, for the 100 cards you can pull with your Bayes Bucks, but this gets really tedious, so instead we use the special mathematical notation Σ (the capital Greek letter sigma):

$$\sum_{k=1}^{100} \binom{100}{k} \times 0.00720^k \times (1 - 0.00720)^{n-k}$$

The Σ is the summation symbol; the number at the bottom represents the value we start with and the number at the top represents the value we end with. So the preceding equation is simply adding up the values for the binomial distribution for every value of k from 1 to n, for a p of 0.00720.

We've made writing this problem down much easier, but now we actually need to compute this value. Rather than pulling out your calculator to solve this problem, now is a great time to start using R. In R, we can use the pbinom() function to automatically sum up all these values for k in our PMF. Figure 4-3 shows how we use pbinom() to solve our specific problem.

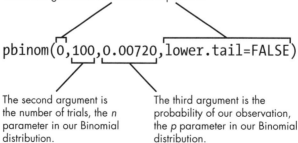

When lower.tail is FALSE, we are looking at the sum of values greater than our k argument. When it is TRUE (or left out), we are looking at values less than or equal to k.

pbinom(0,100,0.00720,lower.tail=FALSE)

The second argument is the number of trials, the n parameter in our Binomial distribution.

The third argument is the probability of our observation, the p parameter in our Binomial distribution.

Figure 4-3: Using the pbinom() function to solve our Bayesian Battlers problem

The `pbinom()` function takes three required arguments and an optional fourth called `lower.tail` (which defaults to `TRUE`). When the fourth argument is `TRUE`, the first argument sums up all of the probabilities *less than or equal* to our argument. When `lower.tail` is set to `FALSE`, it sums up the probabilities *strictly greater than* the first argument. By setting the first argument to 0, we are looking at the probability of getting one or more E. T. Jaynes cards. We set `lower.tail` to `FALSE` because that means we want values greater than the first argument (by default, we get values less than the first argument). The next value represents n, the number of trials, and the third argument represents p, the probability of success.

If we plug in our numbers here and set `lower.tail` to `FALSE` as shown in Figure 4-3, R will calculate your probability of getting *at least one* E. T. Jaynes card for your 100 Bayes Bucks:

$$\sum_{k=1}^{100} \binom{100}{k} \times 0.00720^k \times \left(1 - p\right)^{n-k} = 0.515$$

Even though the probability of getting *exactly one* E. T. Jaynes card is only 0.352, the probability of getting *at least one* E. T. Jaynes card is high enough for you to risk it. So shell out that $10 and complete your set of elite Bayesians!

Wrapping Up

In this chapter we saw that we can use our rules of probability (combined with a trick from combinatorics) to create a general rule that solves an entire class of problems. Any problem that involves wanting to determine the probability of k outcomes in n trials, where the probability of the outcomes is p, we can solve easily using the binomial distribution:

$$B\left(k; n, p\right) = \binom{n}{k} \times p^k \times \left(1 - p\right)^{n-k}$$

Perhaps surprisingly, there is nothing more to this rule than counting and applying our rules of probability.

Exercises

Try answering the following questions to make sure you've grasped binomial distributions fully. The solutions can be found at *https://nostarch.com/learnbayes/*.

1. What are the parameters of the binomial distribution for the probability of rolling either a 1 or a 20 on a 20-sided die, if we roll the die 12 times?

2. There are four aces in a deck of 52 cards. If you pull a card, return the card, then reshuffle and pull a card again, how many ways can you pull just one ace in five pulls?

3. For the example in question 2, what is the probability of pulling five aces in 10 pulls (remember the card is shuffled back in the deck when it is pulled)?

4. When you're searching for a new job, it's always helpful to have more than one offer on the table so you can use it in negotiations. If you have a 1/5 probability of receiving a job offer when you interview, and you interview with seven companies in a month, what is the probability you'll have at least two competing offers by the end of that month?

5. You get a bunch of recruiter emails and find out you have 25 interviews lined up in the next month. Unfortunately, you know this will leave you exhausted, and the probability of getting an offer will drop to 1/10 if you're tired. You really don't want to go on this many interviews unless you are at least twice as likely to get at least two competing offers. Are you more likely to get at least two offers if you go for 25 interviews, or stick to just 7?

5

THE BETA DISTRIBUTION

This chapter builds on the ideas behind the binomial distribution from the previous chapter to introduce another probability distribution, the *beta distribution.* You use the beta distribution to estimate the probability of an event for which you've already observed a number of trials and the number of successful outcomes. For example, you would use it to estimate the probability of flipping a heads when so far you have observed 100 tosses of a coin and 40 of those were heads.

While exploring the beta distribution, we'll also look at the differences between probability and statistics. Often in probability texts, we are given the probabilities for events explicitly. However, in real life, this is rarely the case. Instead, we are given data, which we use to come up with estimates for probabilities. This is where statistics comes in: it allows us to take data and make estimates about what probabilities we're dealing with.

A Strange Scenario: Getting the Data

Here's the scenario for this chapter. One day you walk into a curiosity shop. The owner greets you and, after you browse for a bit, asks if there is anything in particular you're looking for. You respond that you'd love to see the strangest thing he has to show you. He smiles and pulls something out from behind the counter. You're handed a black box, about the size of a Rubik's Cube, that seems impossibly heavy. Intrigued, you ask, "What does it do?"

The owner points out a small slit on the top of the box and another on the bottom. "If you put a quarter in the top," he tells you, "sometimes two come out the bottom!" Excited to try this out, you grab a quarter from your pocket and put it in. You wait and nothing happens. Then the shop owner says, "And sometimes it just eats your quarter. I've had this thing a while, and I've never seen it run out of quarters or get too full to take more!"

Perplexed by this but eager to make use of your newfound probability skills, you ask, "What's the probability of getting two quarters?" The owner replies quizzically, "I have no idea. As you can see, it's just a black box, and there are no instructions. All I know is how it behaves. Sometimes you get two quarters back, and sometimes it eats your quarter."

Distinguishing Probability, Statistics, and Inference

While this is a somewhat unusual everyday problem, it's actually an extremely common type of probability problem. In all of the examples so far, outside of the first chapter, we've known the probability of all the possible events, or at least how much we'd be willing to bet on them. In real life we are almost never sure what the exact probability of any event is; instead, we just have observations and data.

This is commonly considered the division between probability and statistics. In probability, we know exactly how probable all of our events are, and what we are concerned with is how likely certain observations are. For example, we might be told that there is $1/2$ probability of getting heads in a fair coin toss and want to know the probability of getting exactly 7 heads in 20 coin tosses.

In statistics, we would look at this problem backward: assuming you observe 7 heads in 20 coin tosses, what is the probability of getting heads in a single coin toss? As you can see, in this example we don't know what the probability is. In a sense, statistics is probability in reverse. The task of figuring out probabilities given data is called *inference*, and it is the foundation of statistics.

Collecting Data

The heart of statistical inference is data! So far we have only a single sample from the strange box: you put in a quarter and got nothing back. All we know at this point is that it's possible to lose your money. The shopkeeper said you can win, but we don't know that for sure yet.

We want to estimate the probability that the mysterious box will deliver two quarters, and to do that, we first need to see how frequently you win after a few more tries.

The shopkeeper informs you that he's just as curious as you are and will gladly donate a roll of quarters—containing $10 worth of quarters, or 40 quarters—provided you return any winnings to him. You put a quarter in, and happily, two more quarters pop out! Now we have two pieces of data: the mystical box does in fact pay out sometimes, and sometimes it eats the coin.

Given our two observations, one where you lose the quarter and another where you win, you might guess naively that P(two quarters) = 1/2. Since our data is so limited, however, there is still a range of probabilities we might consider for the true rate at which this mysterious box returns two coins. To gather more data, you'll use the rest of the quarters in the roll. In the end, including your first quarter, you get:

14 wins

27 losses

Without doing any further analysis, you might intuitively want to update your guess that P(two quarters) = 1/2 to P(two quarters) = 14/41. But what about your original guess—does your new data mean it's impossible that 1/2 is the real probability?

Calculating the Probability of Probabilities

To help solve this problem, let's look at our two possible probabilities. These are just our hypotheses about the rate at which the magic box returns two quarters:

$$P\left(\text{two coins}\right) = \frac{1}{2} \text{ vs. } P\left(\text{two coins}\right) = \frac{14}{41}$$

To simplify, we'll assign each hypothesis a variable:

$$H_1 \text{ is } P\left(\text{two coins}\right) = \frac{1}{2}$$

$$H_2 \text{ is } P\left(\text{two coins}\right) = \frac{14}{41}$$

Intuitively, most people would say that H_2 is more likely because this is exactly what we observed, but we need to demonstrate this mathematically to be sure.

We can think of this problem in terms of how well each hypothesis explains what we saw, so in plain English: "How probable is what we observed if H_1 were true versus if H_2 were true?" As it turns out, we can easily calculate this using the binomial distribution from Chapter 4. In this case, we know that $n = 41$ and $k = 14$, and for now, we'll assume that $p = H_1$ or H_2. We'll use

D as a variable for our data. When we plug these numbers into the binomial distribution, we get the following results (recall that you can do this with the formula for the binomial distribution in Chapter 4):

$$P(D \mid H_1) = B\left(14;41,\frac{1}{2}\right) \approx 0.016$$

$$P(D \mid H_2) = B\left(14;41,\frac{14}{41}\right) \approx 0.130$$

In other words, if H_1 were true and the probability of getting two coins was 1/2, then the probability of observing 14 occasions where we get two coins out of 41 trials would be about 0.016. However, if H_2 were true and the real probability of getting two coins out of the box was 14/41, then the probability of observing the same outcomes would be about 0.130.

This shows us that, given the data (observing 14 cases of getting two coins out of 41 trials), H_2 is almost 10 times more probable than H_1! However, it also shows that neither hypothesis is *impossible* and that there are, of course, many other hypotheses we could make based on our data. For example, we might read our data as H_3 P(two coins) = 15/42. If we wanted to look for a pattern, we could also pick every probability from 0.1 to 0.9, incrementing by 0.1; calculate the probability of the observed data in each distribution; and develop our hypothesis from that. Figure 5-1 illustrates what each value looks like in the latter case.

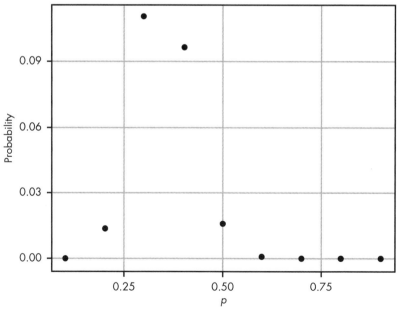

Figure 5-1: Visualization of different hypotheses about the rate of getting two quarters

Even with all these hypotheses, there's no way we could cover every possible eventuality because we're not working with a finite number of hypotheses. So let's try to get more information by testing more distributions. If we repeat the last experiment, testing each possibility at certain increments starting with 0.01 and ending with 0.99, incrementing by only 0.01 would give us the results in Figure 5-2.

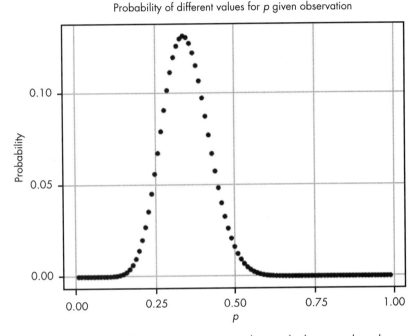

Figure 5-2: We see a definite pattern emerging when we look at more hypotheses.

We may not be able to test every possible hypothesis, but it's clear a pattern is emerging here: we see something that looks like a distribution representing what we believe is the behavior of the black box.

This seems like valuable information; we can easily see where the probability is highest. Our goal, however, is to model our beliefs in all possible hypotheses (that is, the full probability distribution of our beliefs). There are still two problems with our approach. First, because there's an infinite number of possible hypotheses, incrementing by smaller and smaller amounts doesn't accurately represent the entire range of possibilities—we're always missing an infinite amount. In practice, this isn't a huge problem because we often don't care about the extremes like 0.000001 and 0.0000011, but the data would be more useful if we could represent this infinite range of possibilities a bit more accurately.

Second, if you looked at the graph closely, you may have noticed a larger problem here: there are at least 10 dots above 0.1 right now, and we have an infinite number of points to add. This means that our probabilities *don't sum to 1*! From the rules of probability, we know that the probabilities of all our possible hypotheses must sum to 1. If they don't, it means that some hypotheses are not covered. If they add up to more than 1, we would be violating the rule that probabilities must be between 0 and 1. Even though there are infinitely many possibilities here, we still need them all to sum to 1. This is where the beta distribution comes in.

The Beta Distribution

To solve both of these problems, we'll be using the beta distribution. Unlike the binomial distribution, which breaks up nicely into discrete values, the beta distribution represents a continuous range of values, which allows us to represent our infinite number of possible hypotheses.

We define the beta distribution with a *probability density function (PDF)*, which is very similar to the probability mass function we use in the binomial distribution, but is defined for continuous values. Here is the formula for the PDF of the beta distribution:

$$\text{Beta}(p;\alpha,\beta) = \frac{p^{\alpha-1} \times (1-p)^{\beta-1}}{\text{beta}(\alpha,\beta)}$$

Now this looks like a much more terrifying formula than the one for our binomial distribution! But it's actually not that different. We won't build this formula entirely from scratch like we did with the probability mass function, but let's break down some of what's happening here.

Breaking Down the Probability Density Function

Let's first take a look at our parameters: p, α (lowercase Greek letter alpha), and β (lowercase Greek letter beta).

p Represents the probability of an event. This corresponds to our different hypotheses for the possible probabilities for our black box.

α Represents how many times we observe an event we care about, such as getting two quarters from the box.

β Represents how many times the event we care about *didn't* happen. For our example, this is the number of times that the black box ate the quarter.

The total number of trials is $\alpha + \beta$. This is different than the binomial distribution, where we have k observations we're interested in and a finite number of n total trials.

The top part of the PDF function should look pretty familiar because it's almost the same as the binomial distribution's PMF, which looks like this:

$$B(k;n,p) = \binom{n}{k} \times p^{k} \times (1-p)^{n-k}$$

In the PDF, rather than $p^k \times (1-p)^{n-k}$, we have $p^{\alpha-1} \times (1-p)^{\beta-1}$ where we subtract 1 from the exponent terms. We also have another function in the denominator of our equation: the *beta* function (note the lowercase) for which the beta distribution is named. We subtract 1 from the exponent and use the beta function to *normalize* our values—this is the part that ensures our distribution sums to 1. The beta function is the *integral* from 0 to 1 of $p^{\alpha-1} \times (1-p)^{\beta-1}$. We'll talk about integrals more in the next section, but you can think of this as the sum of all the possible values of $p^{\alpha-1} \times (1-p)^{\beta-1}$ when p is every number between 0 and 1. A discussion of how subtracting 1 from the exponents and dividing by the beta functions normalizes our values is beyond the scope of this chapter; for now, you just need to know that this allows our values to sum to 1, giving us a workable probability.

What we get in the end is a function that describes the probability of each possible hypothesis for our true belief in the probability of getting two heads from the box, given that we have observed α examples of one outcome and β examples of another. Remember that we arrived at the beta distribution by comparing how well different binomial distributions, each with its own probability p, described our data. In other words, the beta distribution represents how well all possible binomial distributions describe the data observed.

Applying the Probability Density Function to Our Problem

When we plug in our values for our black box data and visualize the beta distribution, shown in Figure 5-3, we see that it looks like a smooth version of the plot in Figure 5-2. This illustrates the PDF of Beta(14,27).

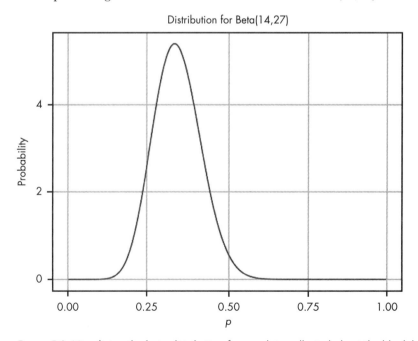

Figure 5-3: Visualizing the beta distribution for our data collected about the black box

As you can see, most of the plot's density is less than 0.5, as we would expect given that our data shows that fewer than half of the quarters placed in the black box returned two quarters.

The plot also shows that it's very unlikely the black box will return two quarters at least half the time, which is the point at which we break even if we continually put quarters in the box. We've figured out that we're more likely to lose money than make money through the box, without sacrificing too many quarters. While we can see the distribution of our beliefs by looking at a plot, we'd still like to be able to quantify exactly how strongly we believe that "the probability that the true rate at which the box returns two quarters is less than 0.5." To do this, we need just a bit of calculus (and some R).

Quantifying Continuous Distributions with Integration

The beta distribution is fundamentally different from the binomial distribution in that with the latter, we are looking at the distribution of k, the number of outcomes we care about, which is always something we can count. For the beta distribution, however, we are looking at the distribution of p, for which we have an infinite number of possible values. This leads to an interesting problem that might be familiar if you've studied calculus before (but it's okay if you haven't!). For our example of $\alpha=14$ and $\beta=27$, we want to know: what is the probability that the chance of getting two coins is 1/2?

While it's easy to ask the likelihood of an exact value with the binomial distribution thanks to its finite number of outcomes, this is a really tricky question for a continuous distribution. We know that the fundamental rule of probability is that the sum of all our values must be 1, but each of our individual values is *infinitely* small, meaning the probability of any specific value is in practice 0.

This may seem strange if you aren't familiar with continuous functions from calculus, so as a quick explanation: this is just the logical consequence of having something made up of an infinite number of pieces. Imagine, for example, you divide a 1-pound bar of chocolate (pretty big!) into two pieces. Each piece would then weigh 1/2 a pound. If you divided it into 10 pieces, each piece would weigh 1/10 a pound. As the number of pieces you divide the chocolate into grows, each piece becomes so small you can't even see it. For the case where the number of pieces goes to infinity, eventually those pieces disappear!

Even though the individual pieces disappear, we can still talk about ranges. For example, even if we divided a 1-pound bar of chocolate into infinitely many pieces, we can still add up the weight of the pieces in one half of the chocolate bar. Similarly, when talking about probability in continuous distributions, we can sum up ranges of values. But if every specific value is 0, then isn't the sum just 0 as well?

This is where calculus comes in: in calculus, there's a special way of summing up infinitely small values called the *integral*. If we want to know whether the probability that the box will return a coin is less than 0.5 (that is, the value is somewhere between 0 and 0.5), we can sum it up like this:

$$\int_0^{0.5} \frac{p^{14-1} \times (1-p)^{27-1}}{\text{beta}(14,27)}$$

If you're rusty on calculus, the stretched-out *S* is the continuous function equivalent to Σ for discrete functions. It's just a way to express that we want to add up all the little bits of our function (see Appendix B for a quick overview of the basic principles of calculus).

If this math is starting to look too scary, don't worry! We'll use R to calculate this for us. R includes a function called dbeta() that is the PDF for the beta distribution. This function takes three arguments, corresponding to *p*, α, and β. We use this together with R's integrate() function to perform this integration automatically. Here we calculate the probability that the chance of getting two coins from the box is 0.5, given the data:

```
> integrate(function(p) dbeta(p,14,27),0,0.5)
```

The result is as follows:

```
0.9807613 with absolute error < 5.9e-06
```

The "absolute error" message appears because computers can't perfectly calculate integrals so there is always some error, though usually it is far too small for us to worry about. This result from R tells us that there is a 0.98 probability that, given our evidence, the true probability of getting two coins out of the black box is less than 0.5. This means it would not be good idea to put any more quarters in the box, since you very likely won't break even.

Reverse-Engineering the Gacha Game

In real-life situations, we almost never know the true probabilities for events. That's why the beta distribution is one of our most powerful tools for understanding our data. In the Gacha game in Chapter 4, we knew the probability of each card we wanted to pull. In reality, the game developers are very unlikely to give players this information, for many reasons (such as not wanting players to calculate how unlikely they are to get the card they want). Now suppose we are playing a new Gacha game called *Frequentist Fighters!* and it also features famous statisticians. This time, we are pulling for the Bradley Efron card.

We don't know the rates for the card, but we really want that card—and more than one if possible. We spend a ridiculous amount of money and find that from 1,200 cards pulled, we received only 5 Bradley Efron cards. Our friend is thinking of spending money on the game but only wants to do it if there is a better than 0.7 probability that the chance of pulling a Bradley Efron is greater than 0.005.

Our friend has asked us to figure out whether he should spend the money and pull. Our data tells us that of 1,200 cards pulled, only 5 were Bradley Efron, so we can visualize this as Beta(5,1195), shown in Figure 5-4 (remember that the total cards pulled is $\alpha + \beta$).

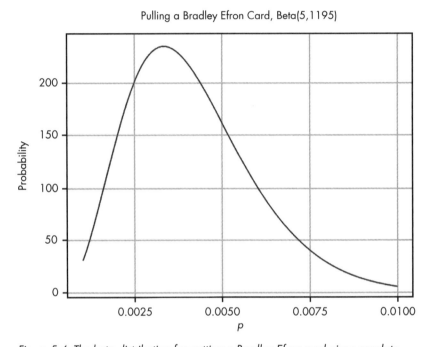

Figure 5-4: The beta distribution for getting a Bradley Efron card given our data

From our visualization we can see that nearly all the probability density is below 0.01. We need to know exactly how much is above 0.005, the value that our friend cares about. We can solve this by integrating over the beta distribution in R, as earlier:

```
integrate(function(x) dbeta(x,5,1195),0.005,1)
0.29
```

This tells us the probability that the rate of pulling a Bradley Efron card is 0.005 or greater, given the evidence we have observed, is only 0.29. Our friend will pull for this card only if the probability is around 0.7 or greater, so based on the evidence from our data collection, our friend should *not* try his luck.

Wrapping Up

In this chapter, you learned about the beta distribution, which is closely related to the binomial distribution but behaves quite differently. We built up to the beta distribution by observing how well an increasing number of possible binomial distributions explained our data. Because our number of possible hypotheses was infinite, we needed a continuous probability distribution that could describe all of them. The beta distribution allows us to represent how strongly we believe in all possible probabilities for the data we observed. This enables us to perform statistical inference on observed data by determining which probabilities we might assign to an event and how strongly we believe in each one: a probability of probabilities.

The major difference between the beta distribution and the binomial distribution is that the beta distribution is a *continuous* probability distribution. Because there are an infinite number of values in the distribution, we cannot sum results the same way we do in a discrete probability distribution. Instead, we need to use calculus to sum ranges of values. Fortunately, we can use R instead of solving tricky integrals by hand.

Exercises

Try answering the following questions to make sure you understand how we can use the Beta distribution to estimate probabilities. The solutions can be found at *https://nostarch.com/learnbayes/*.

1. You want to use the beta distribution to determine whether or not a coin you have is a fair coin—meaning that the coin gives you heads and tails equally. You flip the coin 10 times and get 4 heads and 6 tails. Using the beta distribution, what is the probability that the coin will land on heads more than 60 percent of the time?

2. You flip the coin 10 more times and now have 9 heads and 11 tails total. What is the probability that the coin is fair, using our definition of fair, give or take 5 percent?

3. Data is the best way to become more confident in your assertions. You flip the coin 200 more times and end up with 109 heads and 111 tails. Now what is the probability that the coin is fair, give or take 5 percent?

PART II

BAYESIAN PROBABILITY AND PRIOR PROBABILITIES

6

CONDITIONAL PROBABILITY

So far, we have dealt only with *independent* probabilities. Probabilities are independent when the outcome of one event does not affect the outcome of another. For example, flipping heads on a coin doesn't impact whether or not a die will roll a 6. Calculating probabilities that are independent is much easier than calculating probabilities that aren't, but independent probabilities often don't reflect real life. For example, the probability that your alarm doesn't go off and the probability that you're late for work are *not* independent. If your alarm doesn't go off, you are far more likely to be late for work than you would otherwise be.

In this chapter, you'll learn how to reason about *conditional* probabilities, where probabilities are not independent but rather depend on the outcome of particular events. I'll also introduce you to one of the most important applications of conditional probability: Bayes' theorem.

Introducing Conditional Probability

In our first example of conditional probabilities, we'll look at flu vaccines and possible complications of receiving them. When you get a flu vaccine, you're typically handed a sheet of paper that informs you of the various risks associated with it. One example is an increased incidence of Guillain-Barré syndrome (GBS), a very rare condition that causes the body's immune system to attack the nervous system, leading to potentially life-threatening complications. According to the Centers for Disease Control and Prevention (CDC), the probability of contracting GBS in a given year is 2 in 100,000. We can represent this probability as follows:

$$P(\text{GBS}) = \frac{2}{100,000}$$

Normally the flu vaccine increases your probability of getting GBS only by a trivial amount. In 2010, however, there was an outbreak of swine flu, and the probability of getting GBS if you received the flu vaccine that year rose to 3/100,000. In this case, the probability of contracting GBS directly depended on whether or not you got the flu vaccine, and thus it is an example of a conditional probability. We express conditional probabilities as $P(A \mid B)$, or *the probability of A given B*. Mathematically, we can express the chance of getting GBS as:

$$P(\text{GBS} \mid \text{flu vaccine}) = \frac{3}{100,000}$$

We read this expression in English as "The probability of having GBS, given that you got the flu vaccine, is 3 in 100,000."

Why Conditional Probabilities Are Important

Conditional probabilities are an essential part of statistics because they allow us to demonstrate how information changes our beliefs. In the flu vaccine example, if you don't know whether or not someone got the vaccine, you can say that their probability of getting GBS is 2/100,000 since this is the probability that any given person picked out of the population would have GBS that year. If the year is 2010 and a person tells you that they got the flu shot, you know that the true probability is 3/100,000. We can also look at this as a ratio of these two probabilities, like so:

$$\frac{P(\text{GBS} \mid \text{flu vaccine})}{P(\text{GBS})} = 1.5$$

So if you had the flu shot in 2010, we have enough information to believe you're 50 percent more likely to get GBS than a stranger picked at random. Fortunately, on an individual level, the probability of getting GBS is still very low. But if we're looking at populations as a whole, we would expect 50 percent more people to have GBS in a population of people that had the flu vaccine than in the general population.

There are also other factors that can increase the probability of getting GBS. For example, males and older adults are more likely to have GBS. Using conditional probabilities, we can add all of this information to better estimate the likelihood that an individual gets GBS.

Dependence and the Revised Rules of Probability

As a second example of conditional probabilities, we'll use color blindness, a vision deficiency that makes it difficult for people to discern certain colors. In the general population, about 4.25 percent of people are color blind. The vast majority of cases of color blindness are genetic. Color blindness is caused by a defective gene in the X chromosome. Because males have only a single X chromosome and females have two, men are about 16 times more likely to suffer adverse effects of a defective X chromosome and therefore to be color blind. So while the rate of color blindness for the entire population is 4.25 percent, it is only 0.5 percent in females but 8 percent in males. For all of our calculations, we'll be making the simplifying assumption that the male/female split of the population is exactly 50/50. Let's represent these facts as conditional probabilities:

$$P(\text{color blind}) = 0.0425$$
$$P(\text{color blind} \mid \text{female}) = 0.005$$
$$P(\text{color blind} \mid \text{male}) = 0.08$$

Given this information, if we pick a random person from the population, what's the probability that they are male and color blind?

In Chapter 3, we learned how we can combine probabilities with AND using the product rule. According to the product rule, we would expect the result of our question to be:

$$P(\text{male, color blind}) = P(\text{male}) \times P(\text{color blind}) = 0.5 \times 0.0425 = 0.02125$$

But a problem arises when we use the product rule with conditional probabilities. The problem becomes clearer if we try to find the probability that a person is *female* and color blind:

$$P(\text{female, color blind}) = P(\text{female}) \times P(\text{color blind}) = 0.5 \times 0.0425 = 0.02125$$

This can't be right because the two probabilities are the same! We know that, while the probability of picking a male or a female is the same, if we pick a female, the probability that she is color blind should be much lower than for a male. Our formula should account for the fact that if we pick our person at random, then the probability that they are color blind depends on whether they are male or female. The product rule given in Chapter 3 works only when the probabilities are independent. Being male (or female) and color blind are dependent probabilities.

So the true probability of finding a male who is color blind is the probability of picking a male multiplied by the probability that he is color blind. Mathematically, we can write this as:

$$P(\text{male, color blind}) = P(\text{male}) \times P(\text{color blind} \mid \text{male}) = 0.5 \times 0.08 = 0.04$$

We can generalize this solution to rewrite our product rule as follows:

$$P(A, B) = P(A) \times P(B \mid A)$$

This definition works for independent probabilities as well, because for independent probabilities $P(B) = P(B \mid A)$. This makes intuitive sense when you think about flipping heads and rolling a 6; because $P(\text{six})$ is 1/6 independent of the coin toss, $P(\text{six} \mid \text{heads})$ is also 1/6.

We can also update our definition of the sum rule to account for this fact:

$$P(A \text{ or } B) = P(A) + P(B) - P(A) \times P(B \mid A)$$

Now we can still easily use our rules of probabilistic logic from Part I and handle conditional probabilities.

An important thing to note about conditional probabilities and dependence is that, in practice, knowing how two events are related is often difficult. For example, we might ask about the probability of someone owning a pickup truck and having a work commute of over an hour. While we can come up with plenty of reasons one might be dependent on the other—maybe people with pickup trucks tend to live in more rural areas and commute less—we might not have the data to support this. Assuming that two events are independent (even when they likely aren't) is a very common practice in statistics. But, as with our example for picking a color blind male, this assumption can sometimes give us very wrong results. While assuming independence is often a practical necessity, never forget how much of an impact dependence can have.

Conditional Probabilities in Reverse and Bayes' Theorem

One of the most amazing things we can do with conditional probabilities is reversing the condition to calculate the probability of the event we're conditioning on; that is, we can use $P(A \mid B)$ to arrive at $P(B \mid A)$. As an example, say you're emailing a customer service rep at a company that sells color blindness–correcting glasses. The glasses are a little pricey, and you mention to the rep that you're worried they might not work. The rep replies, "I'm also color blind, and I have a pair myself—they work really well!"

We want to figure out the probability that this rep is male. However, the rep provides no information except an ID number. So how can we figure out the probability that the rep is male?

We know that $P(\text{color blind} \mid \text{male}) = 0.08$ and that $P(\text{color blind} \mid \text{female}) = 0.005$, but how can we determine $P(\text{male} \mid \text{color blind})$? Intuitively, we know that it is much more likely that the customer service rep is in fact male, but we need to quantify that to be sure.

Thankfully, we have all the information we need to solve this problem, and we know that we are solving for the probability that someone is male, given that they are color blind:

$$P(\text{male} \mid \text{color blind}) = ?$$

The heart of Bayesian statistics is data, and right now we have only one piece of data (other than our existing probabilities): we know that the customer support rep is color blind. Our next step is to look at the portion of the total population that is color blind; then, we can figure out what portion of that subset is male.

To help reason about this, let's add a new variable N, which represents the total population of people. As stated before, we first need to calculate the total subset of the population that is color blind. We know $P(\text{color blind})$, so we can write this part of the equation like so:

$$P(\text{male} \mid \text{color blind}) = \frac{?}{P(\text{color blind}) \times N}$$

Next we need to calculate the number of people who are male *and* color blind. This is easy to do since we know $P(\text{male})$ and $P(\text{color blind} \mid \text{male})$, and we have our revised product rule. So we can simply multiply this probability by the population:

$$P(\text{male}) \times P(\text{color blind} \mid \text{male}) \times N$$

So the probability that the customer service rep is male, given that they're color blind, is:

$$P(\text{male} \mid \text{color blind}) = \frac{P(\text{male}) \times P(\text{color blind} \mid \text{male}) \times N}{P(\text{color blind}) \times N}$$

Our population variable N is on both the top and the bottom of the fraction, so the Ns cancel out:

$$P(\text{male} \mid \text{color blind}) = \frac{P(\text{male}) \times P(\text{color blind} \mid \text{male})}{P(\text{color blind})}$$

We can now solve our problem since we know each piece of information:

$$P(\text{male} \mid \text{color blind}) = \frac{P(\text{male}) \times P(\text{color blind} \mid \text{male})}{P(\text{color blind})} = \frac{0.5 \times 0.08}{0.0425} = 0.941$$

Given the calculation, we know there is a 94.1 percent chance that the customer service rep is in fact male!

Introducing Bayes' Theorem

There is nothing actually specific to our case of color blindness in the preceding formula, so we should be able to generalize it to any given A and B probabilities. If we do this, we get the most foundational formula in this book, *Bayes' theorem*:

$$P(A \mid B) = \frac{P(A)P(B \mid A)}{P(B)}$$

To understand why Bayes' theorem is so important, let's look at a general form of this problem. Our beliefs describe the world we know, so when we observe something, its conditional probability represents *the likelihood of what we've seen given what we believe*, or:

$$P(\text{observed} \mid \text{belief})$$

For example, suppose you believe in climate change, and therefore you expect that the area where you live will have more droughts than usual over a 10-year period. Your belief is that climate change is taking place, and your observation is the number of droughts in your area; let's say there were 5 droughts in the last 10 years. Determining how likely it is that you'd see exactly 5 droughts in the past 10 years if there *were* climate change during that period may be difficult. One way to do this would be to consult an expert in climate science and ask them the probability of droughts given that their model assumes climate change.

At this point, all you've done is ask, "What is the probability of what I've observed, given that I believe climate change is true?" But what you want is some way to quantify how strongly you believe climate change is really happening, given what you have observed. Bayes' theorem allows you to reverse P(observed | belief), which you asked the climate scientist for, and solve for the likelihood of your beliefs given what you've observed, or:

$$P(\text{belief} \mid \text{observed})$$

In this example, Bayes' theorem allows you to transform your observation of five droughts in a decade into a statement about how strongly you believe in climate change *after* you have observed these droughts. The only other pieces of information you need are the general probability of 5 droughts in 10 years (which could be estimated with historical data) and your initial certainty of your belief in climate change. And while most people would have a different initial probability for climate change, Bayes' theorem allows you to quantify exactly how much the data changes any belief.

For example, if the expert says that 5 droughts in 10 years is very likely if we assume that climate change is happening, most people will change their previous beliefs to favor climate change a little, whether they're skeptical of climate change or they're Al Gore.

However, suppose that the expert told you that in fact, 5 droughts in 10 years was very unlikely given your assumption that climate change is happening. In that case, your prior belief in climate change would weaken slightly given the evidence. The key takeaway here is that Bayes' theorem ultimately allows evidence to change the strength of our beliefs.

Bayes' theorem allows us to take our beliefs about the world, combine them with data, and then transform this combination into an estimate of the strength of our beliefs given the evidence we've observed. Very often our beliefs are just our initial certainty in an idea; this is the $P(A)$ in Bayes' theorem. We often debate topics such as whether gun control will reduce violence, whether increased testing increases student performance, or whether public health care will reduce overall health care costs. But we seldom think about how evidence should change our minds or the minds of those we're debating. Bayes' theorem allows us to observe evidence about these beliefs and quantify *exactly* how much this evidence changes our beliefs.

Later in this book, you'll see how we can compare beliefs as well as cases where data can surprisingly fail to change beliefs (as anyone who has argued with relatives over dinner can attest!).

In the next chapter, we're going to spend a bit more time with Bayes' theorem. We'll derive it once more, but this time with LEGO; that way, we can clearly visualize how it works. We'll also explore how we can understand Bayes' theorem in terms of more specifically modeling our existing beliefs and how data changes them.

Wrapping Up

In this chapter, you learned about conditional probabilities, which are any probability of an event that depends on another event. Conditional probabilities are more complicated to work with than independent probabilities—we had to update our product rule to account for dependencies—but they lead us to Bayes' theorem, which is fundamental to understanding how we can use data to update what we believe about the world.

Exercises

Try answering the following questions to see how well you understand conditional probability and Bayes' theorem. The solutions can be found at *https://nostarch.com/learnbayes/*.

- What piece of information would we need in order to use Bayes' theorem to determine the probability that someone in 2010 who had GBS *also* had the flu vaccine that year?
- What is the probability that a random person picked from the population is female and is *not* color blind?
- What is the probability that a male who received the flu vaccine in 2010 is either color blind or has GBS?

7

BAYES' THEOREM WITH LEGO

In the previous chapter, we covered conditional probability and arrived at a very important idea in probability, Bayes' theorem, which states:

$$P(A \mid B) = \frac{P(B \mid A)P(A)}{P(B)}$$

Notice that here we've made a very small change from Chapter 6, writing $P(B \mid A)P(A)$ instead of $P(A)P(B \mid A)$; the meaning is identical, but sometimes changing the terms around can help clarify different approaches to problems.

With Bayes' theorem, we can reverse conditional probabilities—so when we know the probability $P(B \mid A)$, we can work out $P(A \mid B)$. Bayes' theorem is foundational to statistics because it allows us to go from having the probability of an observation given a belief to determining the strength of that belief given the observation. For example, if we know the probability of sneezing given that you have a cold, we can work backward to determine the probability that you have a cold given that you sneezed. In this way, we use evidence to update our beliefs about the world.

In this chapter, we'll use LEGO to visualize Bayes' theorem and help solidify the mathematics in your mind. To do this, let's pull out some LEGO bricks and put some concrete questions to our equation. Figure 7-1 shows a 6 × 10 area of LEGO bricks; that's a 60-stud area (*studs* are the cylindrical bumps on LEGO bricks that connect them to each other).

Figure 7-1: A 6 × 10-stud LEGO area to help us visualize the space of possible events

We can imagine this as the space of 60 possible, mutually exclusive events. For example, the blue studs could represent 40 students who passed an exam and the red studs 20 students who failed the exam in a class of 60. In the 60-stud area, there are 40 blue studs, so if we put our finger on a random spot, the probably of touching a blue brick is defined like this:

$$P(\text{blue}) = \frac{40}{60} = \frac{2}{3}$$

We would represent the probability of touching a red brick as follows:

$$P(\text{red}) = \frac{20}{60} = \frac{1}{3}$$

The probability of touching either a blue or a red brick, as you would expect, is 1:

$$P(\text{blue}) + P(\text{red}) = 1$$

This means that red and blue bricks alone can describe our entire set of possible events.

Now let's put a yellow brick on top of these two bricks to represent some other possibility—for example, the students that pulled an all-nighter studying and didn't sleep—so it looks like Figure 7-2.

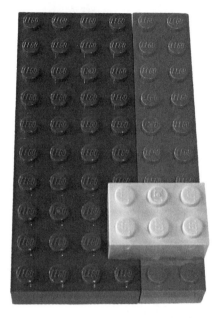

Figure 7-2: Placing a 2 × 3 LEGO brick on top of the 6 × 10-stud LEGO area

Now if we pick a stud at random, the probability of touching the yellow brick is:

$$P(\text{yellow}) = \frac{6}{60} = \frac{1}{10}$$

But if we add $P(\text{yellow})$ to $P(\text{red}) + P(\text{blue})$, we'd get a result greater than 1, and that's impossible!

The issue, of course, is that our yellow studs all sit on top of the space of red and blue studs, so the probability of getting a yellow brick is *conditional* on whether we're on a blue or red space. As we know from the previous chapter, we can express this conditional probability as $P(\text{yellow} \mid \text{red})$, or *the probability of yellow given red*. Given our example from earlier, this would be the probability that a student pulled an all-nighter, given that they had failed an exam.

Working Out Conditional Probabilities Visually

Let's go back to our LEGO bricks and work out $P(\text{yellow} \mid \text{red})$. Figure 7-3 gives us a bit of visual insight into the problem.

Figure 7-3: Visualizing P(yellow | red)

Let's walk through the process for determining $P(\text{yellow} \mid \text{red})$ by working with our physical representation:

1. Split the red section off from the blue.
2. Get the area of the entire red space; it's a 2 × 10-stud area, so that's 20 studs.
3. Get the area of the yellow block on the red space, which is 4 studs.
4. Divide the area of the yellow block by the area of the red block.

This gives us $P(\text{yellow} \mid \text{red}) = 4/20 = 1/5$.

Great—we have arrived at the conditional probability of yellow given red! So far, so good. So what if we now reverse that conditional probability and ask what is $P(\text{red} \mid \text{yellow})$? In plain English, if we know we are on a yellow space, what is the probability that it's red underneath? Or, in our test example, what is the probability that a student failed the exam, given that they pulled an all-nighter?

Looking at Figure 7-3, you may have intuitively figured out $P(\text{red} \mid \text{yellow})$ by reasoning, "There are 6 yellow studs, 4 of which are over red, so the probability of choosing a yellow that's over a red block is 4/6." If you did follow this line of thinking, then congratulations! You just independently discovered Bayes' theorem. But let's quantify that with math to make sure it's right.

Working Through the Math

Getting from our intuition to Bayes' theorem will require a bit of work. Let's begin formalizing our intuition by coming up with a way to *calculate* that there are 6 yellow studs. Our minds arrive at this conclusion through spatial reasoning, but we need to use a mathematical approach. To solve this, we just take the probability of being on a yellow stud multiplied by the total number of studs:

$$\text{numberOfYellowStuds} = P(\text{yellow}) \times \text{totalStuds} = \frac{1}{10} \times 60 = 6$$

The next part of our intuitive reasoning is that 4 of the yellow studs are over red, and this requires a bit more work to prove mathematically. First, we have to establish how many red studs there are; luckily, this is the same process as calculating yellow studs:

$$\text{numberOfRedStuds} = P(\text{red}) \times \text{totalStuds} = \frac{1}{3} \times 60 = 20$$

We've also already figured out the ratio of red studs covered by yellow as $P(\text{yellow}|\text{red})$. To make this a count—rather than a probability—we multiply it by the number of red studs that we just calculated:

$$\text{numberOfRedStuds} = P(\text{yellow}|\text{red}) \times \text{numberOfRedStuds} = \frac{1}{5} \times 20 = 4$$

Finally, we get the ratio of the red studs covered by yellow to the total number of yellow:

$$P(\text{red}|\text{yellow}) = \frac{\text{numberOfRedUnderYellow}}{\text{numberOfYellowStuds}} = \frac{4}{6} = \frac{2}{3}$$

This lines up with our intuitive analysis. However, it doesn't quite look like a Bayes' theorem equation, which should have the following structure:

$$P(A|B) = \frac{P(B|A)P(A)}{P(B)}$$

To get there we'll have to go back and expand the terms in this equation, like so:

$$P(\text{red}|\text{yellow}) = \frac{P(\text{yellow}|\text{red}) \times \text{numberOfRedStuds}}{P(\text{yellow}) \times \text{totalStuds}}$$

We know that we calculate this as follows:

$$P(\text{red}|\text{yellow}) = \frac{P(\text{yellow}|\text{red})P(\text{red}) \times \text{totalStuds}}{P(\text{yellow}) \times \text{totalStuds}}$$

Finally, we just need to cancel out totalStuds from the equation, which gives us:

$$P(\text{red} \mid \text{yellow}) = \frac{P(\text{yellow} \mid \text{red})P(\text{red})}{P(\text{yellow})}$$

From intuition, we have arrived back at Bayes' theorem!

Wrapping Up

Conceptually, Bayes' theorem follows from intuition, but that doesn't mean that the formalization of Bayes' theorem is obvious. The benefit of our mathematical work is that it extracts reason out of intuition. We've confirmed that our original, intuitive beliefs are consistent, and now we have a powerful new tool to deal with problems in probability that are more complicated than LEGO bricks.

In the next chapter, we'll take a look at how to use Bayes' theorem to reason about and update our beliefs using data.

Exercises

Try answering the following questions to see if you have a solid understanding of how we can use Bayes' Theorem to reason about conditional probabilities. The solutions can be found at *https://nostarch.com/learnbayes/*.

1. Kansas City, despite its name, sits on the border of two US states: Missouri and Kansas. The Kansas City metropolitan area consists of 15 counties, 9 in Missouri and 6 in Kansas. The entire state of Kansas has 105 counties and Missouri has 114. Use Bayes' theorem to calculate the probability that a relative who just moved to a county in the Kansas City metropolitan area also lives in a county in Kansas. Make sure to show *P*(Kansas) (assuming your relative either lives in Kansas or Missouri), *P*(Kansas City metropolitan area), and *P*(Kansas City metropolitan area | Kansas).

2. A deck of cards has 52 cards with suits that are either red or black. There are four aces in a deck of cards: two red and two black. You remove a red ace from the deck and shuffle the cards. Your friend pulls a black card. What is the probability that it is an ace?

8

THE PRIOR, LIKELIHOOD, AND POSTERIOR OF BAYES' THEOREM

Now that we've covered how to derive Bayes' theorem using spatial reasoning, let's examine how we can use Bayes' theorem as a probability tool to logically reason about uncertainty. In this chapter, we'll use it to calculate and quantify how likely our belief is, given our data. To do so, we'll use the three parts of the theorem—the posterior probability, likelihood, and prior probability—all of which will come up frequently in your adventures with Bayesian statistics and probability.

The Three Parts

Bayes' theorem allows us to quantify exactly how much our observed data changes our beliefs. In this case, what we want to know is: $P(\text{belief} \mid \text{data})$. In plain English, we want to quantify how strongly we hold our beliefs given the data we've observed. The technical term for this part of the formula is the *posterior probability*, and it's what we'll use Bayes' theorem to solve for.

To solve for the posterior, we need the next part: the probability of the data given our beliefs about the data, or $P(\text{data} \mid \text{belief})$. This is known as the *likelihood*, because it tells us how likely the data is given our belief.

Finally, we want to quantify how likely our initial belief is in the first place, or $P(\text{belief})$. This part of Bayes' theorem is called the *prior probability*, or simply "the prior," because it represents the strength of our belief before we see the data. The likelihood and the prior combine to produce a posterior. Typically we need to use the probability of the data, $P(\text{data})$, in order to normalize our posterior so it accurately reflects a probability from 0 to 1. However, in practice, we don't always need $P(\text{data})$, so this value doesn't have a special name.

As you know by now, we refer to our belief as a hypothesis, H, and we represent our data with the variable D. Figure 8-1 shows each part of Bayes' theorem.

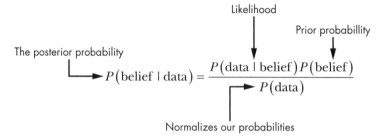

Figure 8-1: The parts of Bayes' theorem

In this chapter, we'll investigate a crime, combining these pieces to reason about the situation.

Investigating the Scene of a Crime

Let's suppose you come home from work one day and find your window broken, your front door open, and your laptop missing. Your first thought is probably "I've been robbed!" But how did you come to this conclusion, and more importantly, how can you quantify this belief?

Your immediate hypothesis is that you have been robbed, so H = I've been robbed. We want a probability that describes how likely it is that you've been robbed, so the posterior we want to solve for given our data is:

$$P(\text{robbed} \mid \text{broken window, open front door, missing laptop})$$

To solve this problem, we'll fill in the missing pieces from Bayes' theorem.

Solving for the Likelihood

First, we need to solve for the likelihood, which in this case is the probability that the same evidence would have been observed if you were in fact robbed—in other words, how closely the evidence lines up with the hypothesis:

$$P(\text{broken window, open front door, missing laptop} \mid \text{robbed})$$

What we're asking is, "If you were robbed, how likely is it that you would see the evidence you saw here?" You can imagine a wide range of scenarios where not all of this evidence was present at a robbery. For example, a clever thief might have picked the lock on your door, stolen your laptop, then locked the door behind them and not needed to break a window. Or they might have just smashed the window, taken the laptop, and then climbed right back out the window. The evidence we've seen seems intuitively like it would be pretty common at the scene of a robbery, so we'll say there's a 3/10 probability that if you were robbed, you would come home and find this evidence.

It's important to note that, even though we're making a guess in this example, we could also do some research to get a better estimate. We could go to the local police department and ask for statistics about evidence at crime scenes involving robbery, or read through news reports of recent robberies. This would give us a more accurate estimate for the likelihood that if you were robbed you would see this evidence.

The incredible thing about Bayes' theorem is that we can use it both for organizing our casual beliefs and for working with large data sets of very exact probabilities. Even if you don't think 3/10 is a good estimate, you can always go back to the calculations—as we will do—and see how the value changes given a different assumption. For example, if you think that the probability of seeing this evidence given a robbery is just 3/100, you can easily go back and plug in those numbers instead. Bayesian statistics lets people disagree about beliefs in a measurable way. Because we are dealing with our beliefs in a quantitative way, you can recalculate everything we do in this chapter to see if this different probability has a substantial impact on any of the final outcomes.

Calculating the Prior

Next, we need to determine the probability that you would get robbed at all. This is our prior. Priors are extremely important, because they allow us to use background information to adjust a likelihood. For example, suppose the scene described earlier happened on a deserted island where you are the only inhabitant. In this case, it would be nearly impossible for you to get robbed (by a human, at least). In another example, if you owned a home in a neighborhood with a high crime rate, robberies might be a frequent occurrence. For simplicity, let's set our prior for being robbed as:

$$P(\text{robbed}) = \frac{1}{1,000}$$

Remember, we can always adjust these figures later given different or additional evidence.

We have nearly everything we need to calculate the posterior; we just need to normalize the data. Before moving on, then, let's look at the unnormalized posterior:

$$P(\text{robbed}) \times P\left(\begin{array}{c}\text{broken window, open front door,}\\ \text{missing laptop} \mid \text{robbed}\end{array}\right) = \frac{3}{10,000}$$

This value is incredibly small, which is surprising since intuition tells us that the probability of your house being robbed given the evidence you observed seems very, very high. But we haven't yet looked at the probability of observing our evidence.

Normalizing the Data

What's missing from our equation is the probability of the data you observed whether or not you were robbed. In our example, this is the probability that you observe that your window is broken, the door is open, and your laptop is missing *all at once*, regardless of the cause. As of now, our equation looks like this:

$$P\left(\begin{array}{c}\text{robbed} \mid \text{broken window,}\\ \text{open front door, missing laptop}\end{array}\right) = \frac{\frac{1}{1,000} \times \frac{3}{10}}{P(D)}$$

The reason the probability in the numerator is so low is that we haven't normalized it with the probability that you would find this strange evidence.

We can see how our posterior changes as we change our $P(D)$ in Table 8-1.

Table 8-1: How the P(D) Affects the Posterior

P(D)	Posterior
0.050	0.006
0.010	0.030
0.005	0.060
0.001	0.300

As the probability of our data decreases, our posterior probability increases. This is because as the data we observe becomes increasingly unlikely, a typically unlikely explanation does a better job of explaining the event (see Figure 8-2).

P(robbed | window,door,laptop)

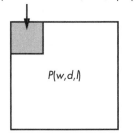

P(w,d,l)

P(robbed | window,door,laptop)

P(w,d,l)

Figure 8-2: As the probability of the data decreases,
the posterior probability increases.

Consider this extreme example: the only way your friend could become
a millionaire is if they won the lottery or inherited money from some fam-
ily member they didn't know existed. Your friend becoming a millionaire
is therefore shockingly unlikely. However, you find out that your friend
did become a millionaire. The possibility that your friend won the lottery
then becomes much more likely, because it is one of the only two ways they
could have become a millionaire.

Being robbed is, of course, only one possible explanation for what you
observed, and there are many more explanations. However, if we don't
know the probability of the evidence, we can't figure out how to normalize
all these other possibilities. So what is our $P(D)$? That's the tricky part.

The common problem with $P(D)$ is that it's very difficult to accurately
calculate in many real-world cases. With every other part of the formula—
even though we just guessed at a value for this exercise—we can collect real
data to provide a more concrete probability. For our prior, $P(\text{robbed})$, we
might simply look at historical crime data and pin down a probability that
a given house on your street would be robbed any given day. Likewise, we
could, theoretically, investigate past robberies and come up with a more
accurate likelihood for observing the evidence you did given a robbery.
But how could we ever really even guess at $P(\text{broken window,open front}$
$\text{door,missing laptop})$?

Instead of researching the probability of the data you observed, we
could try to calculate the probabilities of all other possible events that could
explain your observations. Since they must sum to 1, we could work backward and find $P(D)$. But for the case of this particular evidence, there's a
virtually limitless number of possibilities.

We're a bit stuck without $P(D)$. In Chapters 6 and 7, where we calculated the probability that a customer service rep was male and the probability of choosing different colored LEGO studs, respectively, we had plenty of information about $P(D)$. This allowed us to come up with an exact probability of our belief in our hypothesis given what we observed. Without $P(D)$ we cannot come up with a value for P(robbed | broken window,open front door,missing laptop). However, we're not completely lost.

The good news is that in some cases we don't need to explicitly know $P(D)$, because we often just want to *compare* hypotheses. In this example, we'll compare how likely it is that you were robbed with another possible explanation. We can do this by looking at the ratio of our unnormalized posterior distributions. Because the $P(D)$ would be a constant, we can safely remove it without changing our analysis.

So, instead of calculating the $P(D)$, for the remainder of this chapter we'll develop an alternative hypothesis, calculate its posterior, and then compare it to the posterior from our original hypothesis. While this means we can't come up with an exact probability of being robbed as the only possible explanation for the evidence you observed, we can still use Bayes' theorem to play detective and investigate other possibilities.

Considering Alternative Hypotheses

Let's come up with another hypothesis to compare with our original one. Our new hypothesis consists of three events:

1. A neighborhood kid hit a baseball through the front window.
1. You left your door unlocked.
2. You forgot that you brought your laptop to work and it's still there.

We'll refer to each of these explanations simply by its number in our list, and refer to them collectively as H_2 so that $P(H_2) = P(1,2,3)$. Now we need to solve for the likelihood and prior of this data.

The Likelihood for Our Alternative Hypothesis

Recall that, for our likelihood, we want to calculate the probability of what you observed given our hypothesis, or $P(D | H_2)$. Interestingly—and logically, as you'll see—the likelihood for this explanation turns out to be 1: $P(D | H_2) = 1$

If all the events in our hypothesis did happen, then your observations of a broken window, unlocked door, and missing laptop would be certain.

The Prior for Our Alternative Hypothesis

Our prior represents the possibility of all three events happening. This means we need to first work out the probability of each of these events and then use the product rule to determine the prior. For this example, we'll assume that each of these possible outcomes is conditionally independent.

The first part of our hypothesis is that a neighborhood kid hit a baseball through the front window. While this is common in movies, I've personally never heard of it happening. I have known far more people who have been robbed, though, so let's say that a baseball being hit through the window is half as likely as the probability of getting robbed we used earlier:

$$P(1) = \frac{1}{2,000}$$

The second part of our hypothesis is that you left the door unlocked. This is fairly common; let's say this happens about once a month, so:

$$P(2) = \frac{1}{30}$$

Finally, let's look at leaving your laptop at work. While bringing a laptop to work and leaving it there might be common, completely forgetting you took it in the first place is less common. Maybe this happens about once a year:

$$P(3) = \frac{1}{365}$$

Since we've given each of these pieces of H_2 a probability, we can now calculate our prior probability by applying the product rule:

$$P(H_2) = \frac{1}{2,000} \times \frac{1}{30} \times \frac{1}{365} = \frac{1}{21,900,000}$$

As you can see, the prior probability of all three events happening is extremely low. Now we need a posterior for each of our hypotheses to compare.

The Posterior for Our Alternative Hypothesis

We know that our likelihood, $P(D|H_2)$, equals 1, so if our second hypothesis were to be true, we would be certain to see our evidence. Without a prior probability in our second hypothesis, it looks like the posterior probability for our new hypothesis will be much stronger than it is for our original hypothesis that you were robbed (since we aren't as likely to see the data even if we were robbed). We can now see how the prior radically alters our unnormalized posterior probability:

$$P(D|H_2) \times P(H_2) = 1 \times \frac{1}{21,900,000} = \frac{1}{21,900,000}$$

Now we want to compare our posterior beliefs as well as the strength of our hypotheses with a ratio. You'll see that we don't need a $P(D)$ to do this.

Comparing Our Unnormalized Posteriors

First, we want to compare the ratio of the two posteriors. A ratio tells us how many times more likely one hypothesis is than the other. We'll define our original hypothesis as H_1, and the ratio looks like this:

$$\frac{P(H_1 \mid D)}{P(H_2 \mid D)}$$

Next let's expand this using Bayes' theorem for each of these. We'll write Bayes' theorem as $P(H) \times P(D \mid H) \times 1/P(D)$ to make the formula easier to read in this context:

$$\frac{P(H_1) \times P(D \mid H_1) \times \dfrac{1}{P(D)}}{P(H_2) \times P(D \mid H_2) \times \dfrac{1}{P(D)}}$$

Notice that both the numerator and denominator contain $1/P(D)$, which means we can remove that and maintain the ratio. This is why $P(D)$ doesn't matter when we compare hypotheses. Now we have a ratio of the unnormalized posteriors. Because the posterior tells us how strong our belief is, this ratio of posteriors tells us how many times better H_1 explains our data than H_2 without knowing $P(D)$. Let's cancel out the $P(D)$ and plug in our numbers:

$$\frac{P(H_1) \times P(D \mid H_1)}{P(H_2) \times P(D \mid H_2)} = \frac{\dfrac{3}{10,000}}{\dfrac{1}{21,900,000}} = 6,570$$

What this means is that H_1 explains what we observed 6,570 times better than H_2. In other words, our analysis shows that our original hypothesis (H_1) explains our data much, much better than our alternate hypothesis (H_2). This also aligns well with our intuition—given the scene you observed, a robbery certainly sounds like a more likely assessment.

We'd like to express this property of the unnormalized posterior mathematically to be able to use it for comparison. For that, we use the following version of Bayes' theorem, where the symbol $\propto$ means "proportional to":

$$P(H \mid D) \propto P(H) \times P(D \mid H)$$

We can read this as: "The posterior—that is, the probability of the hypothesis given the data—is *proportional to* the prior probability of H multiplied by the probability of the data given H."

This form of Bayes' theorem is extremely useful whenever we want to compare the probability of two ideas but can't easily calculate $P(D)$. We cannot come up with a meaningful value for the probability of our hypothesis in isolation, but we're still using a version of Bayes' theorem to compare hypotheses. Comparing hypotheses means that we can always see exactly how much stronger one explanation of what we've observed is than another.

Wrapping Up

This chapter explored how Bayes' theorem provides a framework for modeling our beliefs about the world, given data that we have observed. For Bayesian analysis, Bayes' theorem consists of three major parts: the posterior probability, $P(H|D)$; the prior probability, $P(H)$; and the likelihood, $P(D|H)$.

The data itself, or $P(D)$, is notably absent from this list, because we often won't need it to perform our analysis if all we're worried about is comparing beliefs.

Exercises

Try answering the following questions to see if you have a solid understanding of the different parts of Bayes' Theorem. The solutions can be found at *https://nostarch.com/learnbayes/*.

1. As mentioned, you might disagree with the original probability assigned to the likelihood:

$$P\left(\text{broken window, open front door, missing laptop} \mid \text{robbed}\right) = \frac{3}{10}$$

   How much does this change our strength in believing H_1 over H_2?

2. How unlikely would you have to believe being robbed is—our prior for H_1—in order for the ratio of H_1 to H_2 to be even?

9

BAYESIAN PRIORS AND WORKING WITH PROBABILITY DISTRIBUTIONS

Prior probabilities are the most controversial aspect of Bayes' theorem, because they're frequently considered subjective. In practice, however, they often demonstrate how to apply vital background information to fully reason about an uncertain situation.

In this chapter, we'll look at how to use a prior to solve a problem, and at ways to use probability distributions to numerically describe our beliefs as a range of possible values rather than single values. Using probability distributions instead of single values is useful for two major reasons.

First, in reality there is often a wide range of possible beliefs we might have and consider. Second, representing ranges of probabilities allows us to state our confidence in a set of hypotheses. We explored both of these examples when examining the mysterious black box in Chapter 5.

C-3PO's Asteroid Field Doubts

As an example, we'll use one of the most memorable errors in statistical analysis from a scene in *Star Wars: The Empire Strikes Back*. When Han Solo, attempting to evade enemy fighters, flies the *Millennium Falcon* into an asteroid field, the ever-knowledgeable C-3PO informs Han that probability isn't on his side. C-3PO says, "Sir, the possibility of successfully navigating an asteroid field is approximately 3,720 to 1!"

"Never tell me the odds!" replies Han.

Superficially, this is just a fun movie dismissing "boring" data analysis, but there's actually an interesting dilemma here. We the viewers know that Han can pull it off, but we probably also don't disagree with C-3PO's analysis. Even Han believes it's dangerous, saying, "They'd have to be crazy to follow us." Plus, none of the pursuing TIE fighters make it through, which provides pretty strong evidence that C-3PO's numbers aren't totally off.

What C-3PO is missing in his calculations is that Han is a badass! C-3PO isn't wrong, he's just forgetting to add essential information. The question now is: can we find a way to avoid C-3PO's error without dismissing probability entirely, as Han proposes? To answer this question, we need to model both how C-3PO thinks and what we believe about Han, then blend those models using Bayes' theorem.

We'll start with C-3PO's reasoning in the next section, and then we'll capture Han's badassery.

Determining C-3PO's Beliefs

C-3PO isn't just making up numbers. He's fluent in over 6 million forms of communication, and that takes a lot of data to support, so we can assume that he has actual data to back up his claim of "approximately 3,720 to 1." Because C-3PO provides the *approximate* odds of successfully navigating an asteroid field, we know that the data he has gives him only enough information to suggest a range of possible rates of success. To represent that range, we need to look at a *distribution* of beliefs regarding the probability of success, rather than a single value representing the probability.

To C-3PO, the only possible outcomes are successfully navigating the asteroid field or not. We'll determine the various possible probabilities of success, given C-3PO's data, using the beta distribution you learned about in Chapter 5. We're using the beta distribution because it correctly models a range of possible probabilities for an event, given information we have on the rate of successes and failures.

Recall that the beta distribution is parameterized with an α (number of observed successes) and a β (the number of observed failures):

$$P(\text{RateOfSuccess} \mid \text{Successes and Failures}) = \text{Beta}(\alpha, \beta)$$

This distribution tells us which rates of success are most likely given the data we have.

To figure out C-3PO's beliefs, we'll make some assumptions about where his data comes from. Let's say that C-3PO has records of 2 people surviving the asteroid field, and 7,440 people ending their trip in a glorious explosion! Figure 9-1 shows a plot of the probability density function that represents C-3PO's belief in the true rate of success.

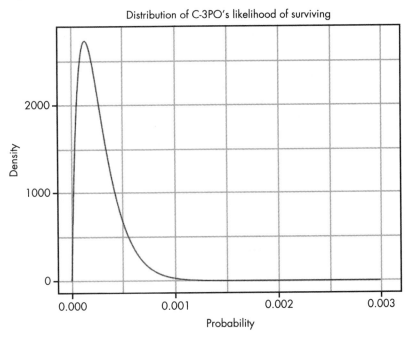

Figure 9-1: A beta distribution representing C-3PO's belief that Han will survive

For any ordinary pilot entering an asteroid field, this looks bad. In Bayesian terms, C-3PO's estimate of the true rate of success given observed data, 3,720:1, is the *likelihood*, which we discussed in Chapter 8. Next, we need to determine our prior.

Accounting for Han's Badassery

The problem with C-3PO's analysis is that his data is on *all* pilots, but Han is far from your average pilot. If we can't put a number to Han's badassery, then our analysis is broken—not just because Han makes it through the asteroid field, but because we *believe* he's going to. Statistics is a tool that aids and organizes our reasoning and beliefs about the world. If our statistical analysis not only contradicts our reasoning and beliefs, but also fails to change them, then something is wrong with our analysis.

We have a *prior belief* that Han will make it through the asteroid field, because Han has survived every improbable situation so far. What makes Han Solo legendary is that no matter how unlikely survival seems, he always succeeds!

The prior probability is often very controversial for data analysts outside of Bayesian analysis. Many people feel that just "making up" a prior is not objective. But this scene is an object chapter in why dismissing our prior beliefs is even more absurd. Imagine watching *Empire* for the first time, getting to this scene, and having a friend sincerely tell you, "Welp, Han is dead now." There's not a chance you'd think it was true. Remember that C-3PO isn't entirely wrong about how unlikely survival is: if your friend said, "Welp, those TIE fighters are dead now," you would likely chuckle in agreement.

Right now, we have many reasons for believing Han will survive, but no numbers to back up that belief. Let's try to put something together.

We'll start with some sort of upper bound on Han's badassery. If we believed Han absolutely could not die, the movie would become predictable and boring. At the other end, our belief that Han will succeed is stronger than C-3PO's belief that he won't, so let's say that our belief that Han will survive is 20,000 to 1.

Figure 9-2 shows the distribution for our prior probability that Han will make it.

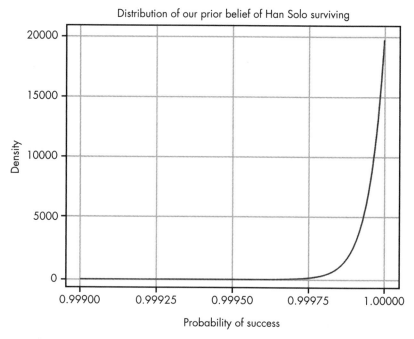

Figure 9-2: The beta distribution representing the range of our prior belief in Han Solo's survival

This is another beta distribution, which we use for two reasons. First, our beliefs are very approximate, so we need to concede a variable rate of survival. Second, a beta distribution will make future calculations much easier.

Now, with our likelihood and prior in hand, we can calculate our posterior probability in the next section.

Creating Suspense with a Posterior

We have now established what C-3PO believes (the likelihood), and we've modeled our own beliefs in Han (the prior), but we need a way to combine these. By combining beliefs, we create our *posterior distribution*. In this case, the posterior models our sense of suspense upon learning the likelihood from C-3PO: the purpose of C-3PO's analysis is in part to poke fun at his analytical thinking, but also to create a sense of real danger. Our prior alone would leave us completely unconcerned for Han, but when we adjust it based on C-3PO's data, we develop a new belief that accounts for the real danger.

The formula for the posterior is actually very simple and intuitive. Given that we have only a likelihood and a prior, we can use the proportional form of Bayes' theorem that we discussed in the previous chapter:

$$\text{Posterior} \propto \text{Likelihood} \times \text{Prior}$$

Remember, using this proportional form of Bayes' theorem means that our posterior distribution doesn't necessarily sum to 1. But we're lucky because there's an easy way to combine beta distributions that will give us a *normalized* posterior when all we have is the likelihood and the prior. Combining our two beta distributions—one representing C-3PO's data (the likelihood) and the other our prior belief in Han's ability to survive anything (our prior)—in this way is remarkably easy:

$$\text{Beta}\left(\alpha_{\text{posterior}}, \beta_{\text{posterior}}\right) = \text{Beta}\left(\alpha_{\text{likelihood}} + \alpha_{\text{prior}}, \beta_{\text{likelihood}} + \beta_{\text{prior}}\right)$$

We just add the alphas for our prior and posterior and the betas for our prior and posterior, and we arrive at a normalized posterior. Because this is so simple, working with the beta distribution is very convenient for Bayesian statistics. To determine our posterior for Han making it through the asteroid field, we can perform this simple calculation:

$$\text{Beta}\left(20002, 7401\right) = \text{Beta}\left(2 + 20000, \ 7400 + 1\right)$$

Now we can visualize our new distribution for our data. Figure 9-3 plots our final posterior belief.

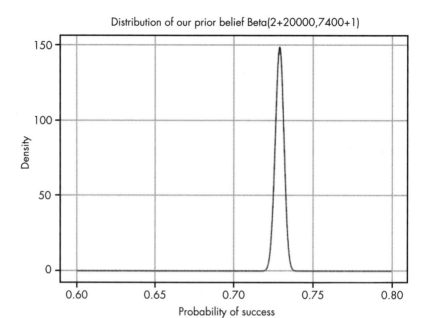

Figure 9-3: Combining our likelihood with our prior gives us a more intriguing posterior.

By combining the C-3PO belief with our Han-is-a-badass belief, we find that we have a far more reasonable position. Our posterior belief is a roughly 73 percent chance of survival, which means we still think Han has a good shot of making it, but we're also still in suspense.

What's really useful is that we don't simply have a raw probability for how likely Han is to make it, but rather a full distribution of possible beliefs. For many examples in the book, we've stuck to simply using a single value for our probabilities, but in practice, using a full distribution helps us to be flexible with the strength of our beliefs.

Wrapping Up

In this chapter, you learned how important background information is to analyzing the data in front of you. C-3PO's data provided us with a likelihood function that didn't match up with our prior understanding of Han's abilities. Rather than simply dismissing C-3PO, as Han famously does, we combine C-3PO's likelihood with our prior to come up with an adjusted belief about the possibility of Han's success. In *Star Wars: The Empire Strikes Back*, this uncertainty is vital for the tension the scene creates. If we completely believe C-3PO's data or our own prior, we would either be nearly certain that Han would die or be nearly certain that he would survive without trouble.

You also saw that you can use probability distributions, rather than a single probability, to express a range of possible beliefs. In later chapters in this book, you'll look at these distributions in more detail to explore the uncertainty of your beliefs in a more nuanced way.

Exercises

Try answering the following questions to see if you understand how to combine prior probability and likelihood distributions to come up with an accurate posterior distribution; solutions to the questions can be found at *https://nostarch.com/learnbayes/.*

1. A friend finds a coin on the ground, flips it, and gets six heads in a row and then one tails. Give the beta distribution that describes this. Use integration to determine the probability that the true rate of flipping heads is between 0.4 and 0.6, reflecting that the coin is reasonably fair.

2. Come up with a prior probability that the coin *is* fair. Use a beta distribution such that there is at least a 95 percent chance that the true rate of flipping heads is between 0.4 and 0.6.

3. Now see how many more heads (with no more tails) it would take to convince you that there is a reasonable chance that the coin is *not* fair. In this case, let's say that this means that our belief in the rate of the coin being between 0.4 and 0.6 drops below 0.5.

PART III

PARAMETER ESTIMATION

10

INTRODUCTION TO AVERAGING AND PARAMETER ESTIMATION

This chapter introduces you to *parameter estimation*, an essential part of statistical inference where we use our data to guess the value of an unknown variable. For example, we might want to estimate the probability of a visitor on a web page making a purchase, the number of jelly beans in a jar at a carnival, or the location and momentum of a particle. In all of these cases, we have an unknown value we want to estimate, and we can use information we have observed to make a guess. We refer to these unknown values as *parameters*, and the process of making the best guess about these parameters as parameter estimation.

We'll focus on *averaging*, which is the most basic form of parameter estimation. Nearly everyone understands that taking an average of a set of observations is the best way to estimate a true value, but few people really stop to ask why this works—if it really does at all. We need to prove that we can trust averaging, because in later chapters, we build it into more complex forms of parameter estimation.

Estimating Snowfall

Imagine there was a heavy snow last night and you'd like to figure out exactly how much snow fell, in inches, in your yard. Unfortunately, you don't have a snow gauge that will give you an accurate measurement. Looking outside, you see that the wind has blown the snow around a bit overnight, meaning it isn't uniformly smooth. You decide to use a ruler to measure the depth at seven roughly random locations in your yard. You come up with the following measurements (in inches):

$$6.2, 4.5, 5.7, 7.6, 5.3, 8.0, 6.9$$

The snow has clearly shifted around quite a bit and your yard isn't perfectly level either, so your measurements are all pretty different. Given that, how can we use these measurements to make a good guess as to the actual snowfall?

This simple problem is a great example case for parameter estimation. The parameter we're estimating is the actual depth of the snowfall from the previous night. Note that, since the wind has blown the snow around and you don't have a snow gauge, we can never know the *exact* amount of snow that fell. Instead, we have a collection of data that we can combine using probability, to determine the contribution of each observation to our estimate, in order to help us make the best possible guess.

Averaging Measurements to Minimize Error

You first instinct is probably to average these measurements. In grade school, we learn to average elements by adding them up and dividing the sum by the total number of elements. So if there are n measurements, each labeled as m_i where i is the ith measurement, we get:

$$\text{average} = \frac{m_1 + m_2 + m_3 \ldots m_n}{n}$$

If we plug in our data, we get the following solution:

$$\frac{(6.2 + 4.5 + 5.7 + 7.6 + 5.3 + 8.0 + 6.9)}{7} = 6.31$$

So, given our seven observations, our best guess is that about 6.31 inches of snow fell.

Averaging is a technique embedded in our minds from childhood, so its application to this problem seems obvious, but in actuality, it's hard to reason about why it works and what it has to do with probability. After all, each of our measurements is different, and all of them are likely different from the true value of the snow that fell. For many centuries, even great mathematicians feared that averaging data compounds all of these erroneous measurements, making for a very inaccurate estimate.

When we estimate parameters, it's vital that we understand *why* we're making a decision; otherwise, we risk using an estimate that may be unintentionally biased or otherwise wrong in a systematic way. One error commonly made in statistics is to blindly apply procedures without understanding them, which frequently leads to applying the wrong solution to a problem. Probability is our tool for reasoning about uncertainty, and parameter estimation is perhaps the most common process for dealing with uncertainty. Let's dive a little deeper into averaging to see if we can become more confident that it is the correct path.

Solving a Simplified Version of Our Problem

Let's simplify our snowfall problem a bit: rather than imagining all possible depths of snow, imagine the snow falling into nice, uniform blocks so that your yard forms a simple two-dimensional grid. Figure 10-1 shows this perfectly even, 6-inch-deep snowfall, visualized from the side (rather than as a bird's-eye view).

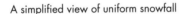

A simplified view of uniform snowfall

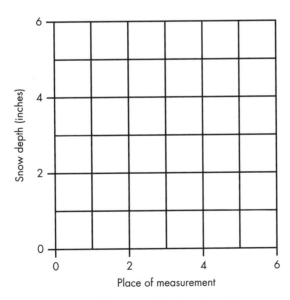

Figure 10-1: Visualizing a perfectly uniform, discrete snowfall

This is the perfect scenario. We don't have an unlimited number of possible measurements; instead, we sample our six possible locations, and each location has only one possible measurement—6 inches. Obviously, averaging works in this case, because no matter how we sample from this data, our answer will always be 6 inches.

Compare that to Figure 10-2, which illustrates the data when we include the windblown snow against the left side of your house.

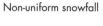
Non-uniform snowfall

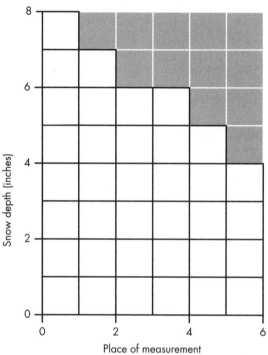

Figure 10-2: Representing the snow shifted by the wind

Now, rather than having a nice, smooth surface, we've introduced some uncertainty into our problem. Of course, we're cheating because we can easily count each block of snow and know exactly how much snow has fallen, but we can use this example to explore how we would reason about an uncertain situation. Let's start investigating our problem by measuring each of the blocks in your yard:

$$8, 7, 6, 6, 5, 4$$

Next, we want to associate some probabilities with each value. Since we're cheating and know the true value of the snowfall is 6 inches, we'll also record the difference between the observation and the true value, known as the *error* value (see Table 10-1).

Table 10-1: Our Observations, and Their Frequencies and Differences from Truth

Observation	Difference from truth	Probability
8	2	1/6
7	1	1/6
6	0	2/6
5	–1	1/6
4	–2	1/6

Looking at the distance from the true measurement for each possible observation, we can see that the probability of overestimating by a certain value is balanced out by the probability of an undervalued measurement. For example, there is a 1/6 probability of picking a measurement that is 2 inches higher than the true value, but there's an equally probable chance of picking a measurement that is 2 inches *lower* than the true measurement. This leads us to our first key insight into why averaging works: errors in measurement tend to cancel each other out.

Solving a More Extreme Case

With such a smooth distribution of errors, the previous scenario might not have convinced you that errors cancel out in more complex situations. To demonstrate how this effect still holds in other cases, let's look at a much more extreme example. Suppose the wind has blown 21 inches of snow to one of the six squares and left only 3 inches at each of the remaining squares, as shown in Figure 10-3.

Now we have a very different distribution of snowfall. For starters, unlike the preceding example, none of the values we can sample from have the true level of snowfall. Also, our errors are no longer nicely distributed—we have a bunch of lower-than-anticipated measurements and one extremely high measurement. Table 10-2 shows the possible measurements, the difference from the true value, and the probability of each measurement.

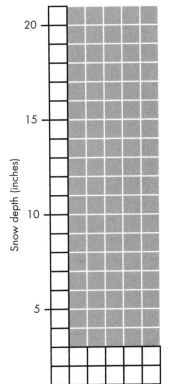

Figure 10-3: A more extreme case of wind shifting the snow

Table 10-2: Observations, Differences, and Probabilities for Our Extreme Example

Observation	Difference from truth	Probability
21	15	1/6
3	-3	5/6

We obviously can't just match up one observation's error value with another's and have them cancel out. However, we can use probability to show that even in this extreme distribution, our errors still cancel each other out. We can do this by thinking of each error measurement as a value that's being voted on by our data. The probability of each error observed is how strongly we believe in that error. When we want to combine our observations, we can consider the probability of the observation as a value representing the strength of its vote toward the final estimate. In this case, the error of –3 inches is five times more likely than the error of 15 inches, so –3 gets weighted more heavily. So, if we were taking a vote, –3 would get five votes, whereas 15 would only get one vote. We combine all of the votes by multiplying each value by its probability and adding them together, giving us a *weighted sum*. In the extreme case where all the values are the same, we would just have 1 multiplied by the value observed and the result would just be that value. In our example, we get:

$$\frac{5}{6} \times -3 + \frac{1}{6} \times 15 = 0$$

The errors in each observation cancel out to 0! So, once again, we find that it doesn't matter if none of the possible values is a true measurement or if the distribution of errors is uneven. When we weight our observations by our belief in that observation, the errors tend to cancel each other out.

Estimating the True Value with Weighted Probabilities

We are now fairly confident that errors from our true measurements cancel out. But we still have a problem: we've been working with the errors from the true observation, but to use these we need to know the true value. When we don't know the true value, all we have to work with are our observations, so we need to see if the errors still cancel out when we have the weighted sum of our original observations.

To demonstrate that our method works, we need some "unknown" true values. Let's start with the following errors:

$$2, 1, -1, -2$$

Since the true measurement is unknown, we'll represent it with the variable t, then add the error. Now we can weight each of these observations by its probability:

$$\frac{1}{4}(2 + t) + \frac{1}{4}(1 + t) + \frac{1}{4}(-1 + t) + \frac{1}{4}(-2 + t)$$

All we've done here is add our error to our constant value *t*, which represents our true measure, then weight each of the results by its probability. We're doing this to see if we can still get our errors to cancel out and leave us with just the value *t*. If so, we can expect errors to cancel out even when we're just averaging raw observations.

Our next step is to apply the probability weight to the values in our terms to get one long summation:

$$\frac{2}{4} + \frac{1}{4}t + \frac{1}{4} + \frac{1}{4}t + \frac{-1}{4} + \frac{1}{4}t + \frac{-2}{4} + \frac{1}{4}t = 0 + t$$

Now if we reorder these terms so that all the errors are together, we can see that our errors will still cancel out, and the weighted *t* value sums up to just *t*, our unknown true value:

$$\left(\frac{2}{4} + \frac{1}{4} + \frac{-1}{4} + \frac{-2}{4}\right) + \left(\frac{1}{4}t + \frac{1}{4}t + \frac{1}{4}t + \frac{1}{4}t\right) = 0 + t$$

This shows that even when we define our measurements as an unknown true value *t* and add some error value, the errors still cancel out! We are left with just the *t* in the end. Even when we don't know what our true measurement or true error is, when we average our values the errors tend to cancel out.

In practice, we typically can't sample the entire space of possible measurements, but the more samples we have, the more the errors are going to cancel out and, in general, the closer our estimate will be to the true value.

Defining Expectation, Mean, and Averaging

What we've arrived at here is formally called the *expectation* or *mean* of our data. It is simply the sum of each value weighted by its probability. If we denote each of our measurements as x_i and the probability of each measurement as p_i, we mathematically define the mean—which is generally represented by μ (the lowercase Greek letter mu)—as follows:

$$\mu = \sum_{1}^{n} p_i x_i$$

To be clear, this is the *exact* same calculation as the averaging we learned in grade school, just with notation to make the use of probability more explicit. As an example, to average four numbers, in school we wrote it as:

$$\frac{x_1 + x_2 + x_3 + x_4}{4}$$

which is identical to writing:

$$\frac{1}{4}x_1 + \frac{1}{4}x_2 + \frac{1}{4}x_3 + \frac{1}{4}x_4$$

or we can just say $p_i = 1/4$ and write it as:

$$\mu = \sum_1^4 p_i x_i$$

So even though the mean is really just the average nearly everyone is familiar with, by building it up from the principles of probability, we see *why* averaging our data works. No matter how the errors are distributed, the probability of errors at one extreme is canceled out by probabilities at the other extreme. As we take more samples, the averages are more likely to cancel out and we start to approach the true measurement we're looking for.

Means for Measurement vs. Means for Summary

We've been using our mean to estimate a true measurement from a distribution of observations with some added error. But the mean is often used as a way to *summarize* a set of data. For example, we might refer to things like:

- The mean height of a person
- The average price of a home
- The average age of a student

In all of these cases, we aren't using mean as a parameter estimate for a single true measurement; instead, we're summarizing the properties of a population. To be precise, we're estimating a parameter of some abstract property of these populations that may not even be real. Even though mean is a very simple and well-known parameter estimate, it can be easily abused and lead to strange results.

A fundamental question you should always ask yourself when averaging data is: "What exactly am I trying to measure and what does this value really mean?" For our snowfall example, the answer is easy: we're trying to estimate how much snow actually fell last night before the wind blew it around. However, when we're measuring the "average height," the answer is less clear. There is no such thing as an average person, and the differences in heights we observe aren't errors—they're truly different heights. A person isn't 5'5" because part of their height drifted onto a 6'3" person!

If you were building an amusement park and wanted to know what height restrictions to put on a roller coaster so that at least half of all visitors could ride it, then you have a real value you are trying to measure. However, in that case, the mean suddenly becomes less helpful. A better measurement to estimate is the probability that someone entering your park will be taller than *x*, where *x* is the minimum height to ride a roller coaster.

All of the claims I've made in this chapter assume we are talking about trying to measure a specific value and using the average to cancel the errors out. That is, we're using averaging as a form of parameter estimation, where our parameter is an actual value that we simply can never know. While averaging can also be useful to summarize large sets of data, we can no longer use the intuition of "errors canceling out" because the variation in the data is genuine, meaningful variation and not error in a measurement.

Wrapping Up

In this chapter, you learned that you can trust your intuition about averaging out your measurements in order to make a best estimate of an unknown value. This is true because errors tend to cancel out. We can formalize this notion of averaging into the idea of the expectation or mean. When we calculate the mean, we are weighting all of our observations by the probability of observing them. Finally, even though averaging is a simple tool to understand, we should always identify and understand what we're trying to determine by averaging; otherwise, our results may end up being invalid.

Exercises

Try answering the following questions to see how well you understand averaging to estimate an unknown measurement. The solutions can be found at *https://nostarch.com/learnbayes/*.

1. It's possible to get errors that don't quite cancel out the way we want. In the Fahrenheit temperature scale, 98.6 degrees is the normal body temperature and 100.4 degrees is the typical threshold for a fever. Say you are taking care of a child that feels warm and seems sick, but you take repeated readings from the thermometer and they all read between 99.5 and 100.0 degrees: warm, but not quite a fever. You try the thermometer yourself and get several readings between 97.5 and 98. What could be wrong with the thermometer?

2. Given that you feel healthy and have traditionally had a very consistently normal temperature, how could you alter the measurements 100, 99.5, 99.6, and 100.2 to estimate if the child has a fever?

11

MEASURING THE SPREAD
OF OUR DATA

In this chapter, you'll learn three different methods—mean absolute deviation, variance, and standard deviation—for quantifying the *spread*, or the different extremes, of your observations.

In the previous chapter, you learned that the mean is the best way to guess the value of an unknown measurement, and that the more spread out our observations, the more uncertain we are about our estimate of the mean. As an example, if we're trying to figure out the location of a collision between two cars based only on the spread of the remaining debris after the cars have been towed away, then the more spread out the debris, the less sure we'd be of where precisely the two cars collided.

Because the spread of our observations is related to the uncertainty in the measurement, we need to be able to quantify it so we can make probabilistic statements about our estimates (which you'll learn how to do in the next chapter).

Dropping Coins in a Well

Say you and a friend are wandering around the woods and stumble across a strange-looking old well. You peer inside and see that it seems to have no bottom. To test it, you pull a coin from your pocket and drop it in, and sure enough, after a few seconds you hear a splash. From this, you conclude that the well is deep, but not bottomless.

With the supernatural discounted, you and your friend are now equally curious as to how deep the well actually is. To gather more data, you grab five more coins from your pocket and drop them in, getting the following measurements in seconds:

$$3.02, 2.95, 2.98, 3.08, 2.97$$

As expected, you find some variation in your results; this is primarily due to the challenge of making sure you drop the coin from the same height and time then record the splash correctly.

Next, your friend wants to try his hand at getting some measurements. Rather than picking five similarly sized coins, he grabs a wider assortment of objects, from small pebbles to twigs. Dropping them in the well, your friend gets the following measurements:

$$3.31, 2.16, 3.02, 3.71, 2.80$$

Both of these samples have a mean (μ) of about 3 seconds, but your measurements and your friend's measurements are spread to different degrees. Our aim in this chapter is to come up with a way to quantify the difference between the spread of your measurements and the spread of your friend's. We'll use this result in the next chapter to determine the probability of certain ranges of values for our estimate.

For the rest of this chapter we'll indicate when we're talking about the first group of values (your observations) with the variable a and the second group (your friend's observations) with the variable b. For each group, each observation is denoted with a subscript; for example, a_2 is the second observation from group a.

Finding the Mean Absolute Deviation

We'll begin by measuring the spread of each observation from the mean (μ). The mean for both a and b is 3. Since μ is our best estimate for the true value, it makes sense to start quantifying the difference in the two spreads by measuring the distance between the mean and each of the values. Table 11-1 displays each observation and its distance from the mean.

Table 11-1: Your and Your Friend's Observations and Their Distances from the Mean

Observation	Difference from mean
Group a	
3.02	0.02
2.95	−0.05
2.98	−0.02
3.08	0.08
2.97	−0.03
Group b	
3.31	0.31
2.16	−0.84
3.02	0.02
3.71	0.71
2.80	−0.16

NOTE   *The distance from the mean is different than the error value, which is the distance from the true value and is unknown in this case.*

A first guess at how to quantify the difference between the two spreads might be to just sum up their differences from the mean. However, when we try this out, we find that the sum of the differences for both sets of observations is exactly the same, which is odd given the notable difference in the spread of the two data sets:

$$\sum_{i=1}^{5} a_1 - \mu_a = 0 \qquad \sum_{i=1}^{5} b_1 - \mu_b = 0$$

The reason we can't simply sum the differences from the mean is related to why the mean works in the first place: as we know from Chapter 10, the errors tend to cancel each other out. What we need is a mathematical method that makes sure our differences don't cancel out without affecting the validity of our measurements.

The reason the differences cancel out is that some are negative and some are positive. So, if we convert all the differences to positives, we can eliminate this problem without invalidating the values.

The most obvious way to do this is to take the *absolute value* of the differences; this is the number's distance from 0, so the absolute value of 4 is 4, and the absolute value of −4 is also 4. This gives us the positive version of our negative numbers without actually changing them. To represent an absolute value, we enclose the value in vertical lines, as in $|{-6}| = |6| = 6$.

If we take the absolute value of the differences in Table 11-1 and use those in our calculation instead, we get a result we can work with:

$$\sum_{1}^{5}\left|a_i - \mu_a\right| = 0.2 \qquad \sum_{1}^{5}\left|b_i - \mu_b\right| = 2.08$$

Try working this out by hand, and you should get the same results. This is a more useful approach for our particular situation, but it applies only when the two sample groups are the same size.

Imagine we had 40 more observations for group a—let's say 20 observations of 2.9 and 20 of 3.1. Even with these additional observations, the data in group a seems less spread out than the data in group b, but the absolute sum of group a is now 85.19 simply because it has more observations!

To correct for this, we can normalize our values by dividing by the total number of observations. Rather than dividing, though, we'll just multiply by 1 over the total, which is known as *multiplying the reciprocal* and looks like this:

$$\frac{1}{5} \times \sum_{1}^{5}\left|a_i - \mu_a\right| = 0.04 \qquad \frac{1}{5} \times \sum_{1}^{5}\left|b_i - \mu_b\right| = 2.08$$

Now we have a measurement of the spread that isn't dependent on the sample size! The generalization of this approach is as follows:

$$\mathrm{MAD}(x) = \frac{1}{n} \times \sum_{1}^{n}\left|x_i - \mu\right|$$

Here we've calculated the mean of the absolute differences between our observations and the mean. This means that for group a the average observation is 0.04 from the mean, and for group b it's about 0.416 seconds from the mean. We call the result of this formula the *mean absolute deviation (MAD)*. The MAD is a very useful and intuitive measure of how spread out your observations are. Given that group a has a MAD of 0.04 and group b around 0.4, we can now say that group b is about 10 times as spread out as group a.

Finding the Variance

Another way to mathematically make all of our differences positive without invalidating the data is to square them: $(x_i - \mu)^2$. This method has at least two benefits over using MAD.

The first benefit is a bit academic: squaring values is much easier to work with mathematically than taking their absolute value. In this book, we won't take advantage of this directly, but for mathematicians, the absolute value function can be a bit annoying in practice.

The second, and more practical, reason is that squaring results in having an *exponential penalty*, meaning measurements very far away from the mean are penalized much more. In other words, small differences aren't nearly as important as big ones, as we would feel intuitively. If someone scheduled your meeting in the wrong room, for example, you wouldn't be too upset if you ended up next door to the right room, but you'd almost certainly be upset if you were sent to an office on the other side of the country.

If we substitute the absolute value for the squared difference, we get the following:

$$\text{Var}(x) = \frac{1}{n} \times \sum_{1}^{n} (x_i - \mu)^2$$

This formula, which has a very special place in the study of probability, is called the *variance*. Notice that the equation for variance is exactly the same as MAD except that the absolute value function in MAD has been replaced with squaring. Because it has nicer mathematical properties, variance is used much more frequently in the study of probability than MAD. We can see how different our results look when we calculate their variance:

Var(group *a*) = 0.002, Var(group *b*) = 0.269

Because we're squaring, however, we no longer have an intuitive understanding of what the results of variance mean. MAD gave us an intuitive definition: this is the average distance from the mean. Variance, on the other hand, says: this is the average squared difference. Recall that when we used MAD, group *b* was about 10 times more spread out than group *a*, but in the case of variance, group *b* is now 100 times more spread out!

Finding the Standard Deviation

While in theory variance has many properties that make it useful, in practice it can be hard to interpret the results. It's difficult for humans to think about what a difference of 0.002 seconds squared means. As we've mentioned, the great thing about MAD is that the result maps quite well to our intuition. If the MAD of group *b* is 0.4, that means that the average distance between any given observation and the mean is literally 0.4 seconds. But averaging over squared differences doesn't allow us to reason about a result as nicely.

To fix this, we can take the square root of the variance in order to scale it back into a number that works with our intuition a bit better. The square root of a variance is called the *standard deviation* and is represented by the lowercase Greek letter sigma (σ). It is defined as follows:

$$\sigma = \sqrt{\frac{1}{n} \times \sum_{1}^{n} (x_i - \mu)^2}$$

The formula for standard deviation isn't as scary as it might seem at first. Looking at all of the different parts, given that our goal is to numerically represent how spread out our data is, we can see that:

1. We want the difference between our data and the mean, $x_i - \mu$.
2. We need to convert negative numbers to positives, so we take the square, $(x_i - \mu)^2$.
3. We need to add up all the differences:

$$\sum_i^n (x_i - \mu)^2$$

4. We don't want the sum to be affected by the number of observations, so we normalize it with $1/n$.
5. Finally, we take the square root of everything so that the numbers are closer to what they would be if we used the more intuitive absolute distance.

If we look at the standard deviation for our two groups, we can see that it's very similar to the MAD:

$$\sigma(\text{group } a) = 0.046, \sigma(\text{group } b) = 0.519$$

The standard deviation is a happy medium between the intuitiveness of MAD and the mathematical ease of variance. Notice that, just like with MAD, the difference in the spread between b and a is a factor of 10. The standard deviation is so useful and ubiquitous that, in most of the literature on probability and statistics, variance is defined simply as σ^2, or sigma squared!

So we now have three different ways of measuring the spread of our data. We can see the results in Table 11-2.

Table 11-2: Measurements of Spread by Method

Method of measuring spread	Group a	Group b
Mean absolute deviations	0.040	0.416
Variance	0.002	0.269
Standard deviation	0.046	0.519

None of these methods for measuring spread is more correct than any other. By far the most commonly used value is the standard deviation, because we can use it, together with the mean, to define a normal distribution, which in turn allows us to define explicit probabilities to possible true values of our measurements. In the next chapter, we'll take a look at the normal distribution and see how it can help us understand our level of confidence in our measurements.

Wrapping Up

In this chapter, you learned three methods for quantifying the spread of a group of observations. The most intuitive measurement of the spread of values is the mean absolute deviation (MAD), which is the average distance of each observation from the mean. While intuitive, MAD isn't as useful mathematically as the other options.

The mathematically preferred method is the variance, which is the squared difference of our observations. But when we calculate the variance, we lose the intuitive feel for what our calculation means.

Our third option is to use the standard deviation, which is the square root of the variance. The standard deviation is mathematically useful and also gives us results that are reasonably intuitive.

Exercises

Try answering the following questions to see how well you understand these different methods of measuring the spread of data. The solutions can be found at *https://nostarch.com/learnbayes/*.

1. One of the benefits of variance is that squaring the differences makes the penalties exponential. Give some examples of when this would be a useful property.

2. Calculate the mean, variance, and standard deviation for the following values: 1, 2, 3, 4, 5, 6, 7, 8, 9, 10.

12

THE NORMAL DISTRIBUTION

In the previous two chapters, you learned about two very important concepts: mean (μ), which allows us to estimate a measurement from various observations, and standard deviation (σ), which allows us to measure the spread of our observations.

On its own, each concept is useful, but together, they are even more powerful: we can use them as parameters for the most famous probability distribution of all, the *normal distribution*.

In this chapter you'll learn how to use the normal distribution to determine an exact probability for your degree of certainty about one estimate proving true compared to others. The true goal of parameter estimation isn't simply to estimate a value, but rather to assign a probability for a *range* of possible values. This allows us to perform more sophisticated reasoning with uncertain values.

We established in the preceding chapter that the mean is a solid method of estimating an unknown value based on existing data, and that the standard deviation can be used to measure the spread of that data. By measuring the spread of our observations, we can determine how confidently we believe in our mean. It makes sense that the more spread out our observations, the less sure we are in our mean. The normal distribution allows us to precisely quantify *how* certain we are in various beliefs when taking our observations into account.

Measuring Fuses for Dastardly Deeds

Imagine a mustachioed cartoon villain wants to set off a bomb to blow a hole in a bank vault. Unfortunately, he has only one bomb, and it's rather large. He knows that if he gets 200 feet away from the bomb, he can escape to safety. It takes him 18 seconds to make it that far. If he's any closer to the bomb, he risks death.

Although the villain has only one bomb, he has six fuses of equal size, so he decides to test out five of the six fuses, saving the last one for the bomb. The fuses are all the same size and should take the same amount of time to burn through. He sets off each fuse and measures how long it takes to burn through to make sure he has the 18 seconds he needs to get away. Of course, being in a rush leads to some inconsistent measurements. Here are the times he recorded (in seconds) for each fuse to burn through: 19, 22, 20, 19, 23.

So far so good: none of the fuses takes less than 18 seconds to burn. Calculating the mean gives us $\mu = 20.6$, and calculating the standard deviation gives us $\sigma = 1.62$.

But now we want to determine a concrete probability for how likely it is that, given the data we have observed, a fuse will go off in *less* than 18 seconds. Since our villain values his life even more than the money, he wants to be 99.9 percent sure he'll survive the blast, or he won't attempt the heist.

In Chapter 10, you learned that the mean is a good estimate for the true value given a set of measurements, but we haven't yet come up with any way to express how *strongly* we believe this value to be true.

In Chapter 11, you learned that you can quantify how spread out your observations are by calculating the standard deviation. It seems rational that this might also help us figure out how likely the alternatives to our mean might be. For example, suppose you drop a glass on the floor and it shatters. When you're cleaning up, you might search adjacent rooms based on how dispersed the pieces of glass are. If, as shown in Figure 12-1, the pieces are very close together, you would feel more confident that you don't need to check for glass in the next room.

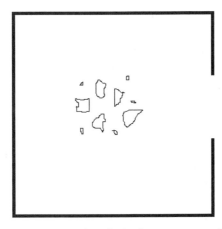

Figure 12-1: When the broken pieces are closer together, you're more sure of where to clean up.

However, if the glass pieces are widely dispersed, as in Figure 12-2, you'll likely want to sweep around the entrance of the next room, even if you don't immediately see broken glass there. Likewise, if the villain's fuse timings are very spread out, even if he didn't observe any fuses lasting less than 18 seconds, it's possible that the real fuse could still burn through in less than 18 seconds.

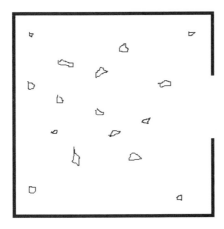

Figure 12-2: When the pieces are spread out, you're less sure of where they might be.

When observations are scattered visually, we intuitively feel that there might be other observations at the extreme limits of what we can see. We are also less confident in exactly where the center is. In the glass example, it's harder to be sure of where the glass fell if you weren't there to witness the fall and the glass fragments are dispersed widely.

We can quantify this intuition with the most studied and well-known probability distribution: the normal distribution.

The Normal Distribution

The normal distribution is a continuous probability distribution (like the beta distribution in Chapter 5) that best describes the strength of possible beliefs in the value of an uncertain measurement, given a known mean and standard deviation. It takes μ and σ (the mean and standard deviation, respectively) as its only two parameters. A normal distribution with $\mu = 0$ and $\sigma = 1$ has a bell shape, as shown in Figure 12-3.

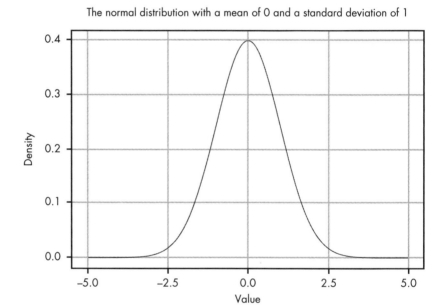

Figure 12-3: A normal distribution with $\mu = 0$ and $\sigma = 1$

As you can see, the center of the normal distribution is its mean. The width of a normal distribution is determined by its standard deviation. Figures 12-4 and 12-5 show normal distributions with $\mu = 0$ and $\sigma = 0.5$ and 2, respectively.

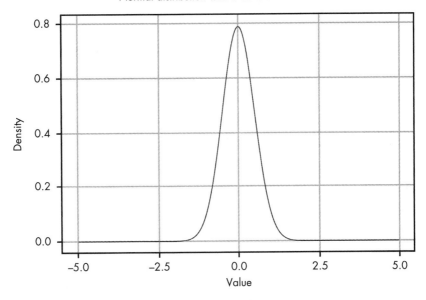

Figure 12-4: A normal distribution with μ = 0 and σ = 0.5

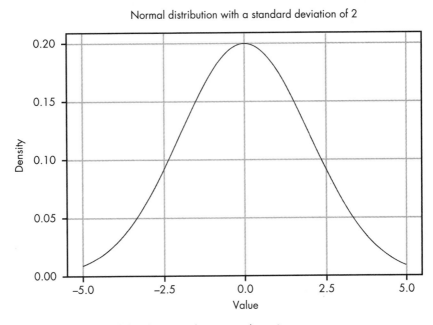

Figure 12-5: A normal distribution with μ = 0 and σ = 2

As the standard deviation shrinks, so does the width of the normal distribution.

The normal distribution, as we've discussed, reflects how strongly we believe in our mean. So, if our observations are more scattered, we believe in a wider range of possible values and have less confidence in the central mean. Conversely, if all of our observations are more or less the same (meaning a small σ), we believe our estimate is pretty accurate.

When the *only* thing we know about a problem is the mean and standard deviation of the data we have observed, the normal distribution is the most honest representation of our state of beliefs.

Solving the Fuse Problem

Going back to our original problem, we have a normal distribution with μ = 20.6 and σ = 1.62. We don't really know anything else about the properties of the fuses beyond the recorded burn times, so we can model the data with a normal distribution using the observed mean and standard deviation (see Figure 12-6).

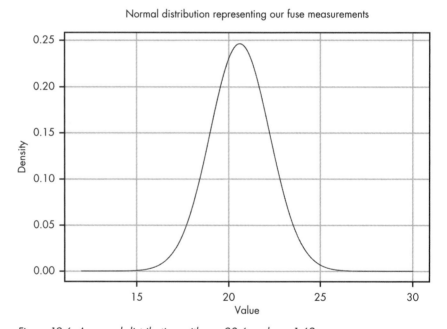

Figure 12-6: A normal distribution with μ = 20.6 and σ = 1.62

The question we want to answer is: what is the probability, given the data observed, that the fuse will run for 18 seconds or less? To solve this problem, we need to use the probability density function (PDF), a concept you first learned about in Chapter 5. The PDF for the normal distribution is:

$$N(\mu,\sigma) = \frac{1}{\sqrt{2\pi\sigma^2}} \times e^{-\frac{(x-\mu)}{2\sigma^2}}$$

And to get the probability, we need to *integrate* this function over values less than 18:

$$\int_{-\infty}^{18} N\left(\mu = 20.6, \sigma = 1.62\right)$$

You can imagine integration as simply taking the area under the curve for the region you're interested in, as shown in Figure 12-7.

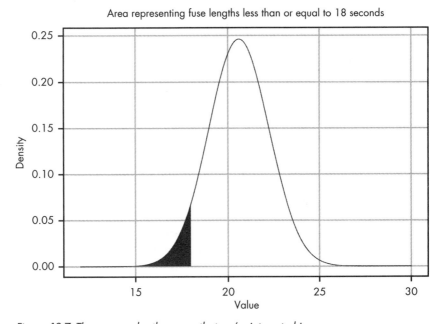

Figure 12-7: The area under the curve that we're interested in

The area of the shaded region represents the probability of the fuse lasting 18 seconds or less given the observations. Notice that even though none of the observed values was less than 18, because of the spread of the observations, the normal distribution in Figure 12-6 shows that a value of 18 or less is still possible. By integrating over all values less than 18, we can calculate the probability that the fuse will *not* last as long as our villain needs it to.

Integrating this function by hand is not an easy task. Thankfully, we have R to do the integration for us.

Before we do this, though, we need to determine what number to start integrating from. The normal distribution is defined on the range of all possible values from negative infinity ($-\infty$) to infinity (∞). So in theory what we want is:

$$P\left(\text{fuse time} < 18\right) = \int_{-\infty}^{18} N\left(\mu, \sigma\right)$$

But obviously we cannot integrate our function from negative infinity on a computer! Luckily, as you can see in Figures 12-6 and 12-7, the probability density function becomes an incredibly small value very quickly. We can see that the line in the PDF is nearly flat at 10, meaning there is virtually no probability in this region, so we can just integrate from 10 to 18. We could also choose a lower value, like 0, but because there's effectively no probability in this region, it won't change our result in any meaningful way. In the next section, we'll discuss a heuristic that makes choosing a lower or upper bound easier.

We'll integrate this function using R's integrate() function and the dnorm() function (which is just R's function for the normal distribution PDF), calculating the PDF of the normal distribution as follows:

```
integrate(function(x) dnorm(x,mean=20.6,sd=1.62),10,18)
0.05425369 with absolute error < 3e-11
```

Rounding the value, we can see that P(fuse time < 18) = 0.05, telling us there is a 5 percent chance that the fuse will last 18 seconds or less. Even villains value their own lives, and in this case our villain will attempt the bank robbery only if he is 99.9 percent sure that he can safely escape the blast. For today then, the bank is safe!

The power of the normal distribution is that we can reason probabilistically about a wide range of possible alternatives to our mean, giving us an idea of how realistic our mean is. We can use the normal distribution any time we want to reason about data for which we know only the mean and standard deviation.

However, this is also the danger of the normal distribution. In practice, if you have information about your problem besides the mean and standard deviation, it is usually best to make use of that. We'll see an example of this in a later section.

Some Tricks and Intuitions

While R makes integrating the normal distribution significantly easier than trying to solve the integral by hand, there's a very useful trick that can simplify things even further when you're working with the normal distribution. For *any* normal distribution with a known mean and standard deviation, you can estimate the area under the curve around μ in terms of σ.

For example, the area under the curve for the range from μ − σ (one standard deviation less than the mean) to μ + σ (one standard deviation greater than the mean) holds 68 percent of the mass of the distribution.

This means that 68 percent of the possible values fall within ± one standard deviation of the mean, as shown in Figure 12-8.

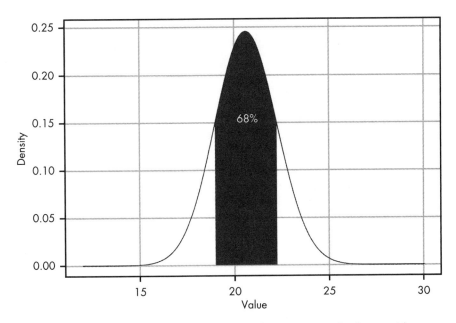

Figure 12-8: Sixty-eight percent of the probability density (area under the curve) lies between one standard deviation of the mean in either direction.

We can continue by increasing our distance from the mean by multiples of σ. Table 12-1 gives probabilities for these other areas.

Table 12-1: Areas Under the Curve for Different Means

Distance from the mean	Probability
σ	68 percent
2σ	95 percent
3σ	99.7 percent

This little trick is very useful for quickly assessing the likelihood of a value given even a small sample. All you need is a calculator to easily figure out the μ and σ, which means you can do some pretty accurate estimations even in the middle of a meeting!

As an example, when measuring snowfall in Chapter 10 we had the following measurements: 6.2, 4.5, 5.7, 7.6, 5.3, 8.0, 6.9. For these measurements, the mean is 6.31 and the standard deviation is 1.17. This means that we can be 95 percent sure that the true value of the snowfall was somewhere between 3.97 inches (6.31 − 2 × 1.17) and 8.65 inches (6.31 + 2 × 1.17). No need to manually calculate an integral or boot up a computer to use R!

Even when we *do* want to use R to integrate, this trick can be useful for determining a minimum or maximum value to integrate from or to. For example, if we want to know the probability that the villain's bomb fuse will last longer than 21 seconds, we don't want to have to integrate from 21 to infinity. What can we use for our upper bound? We can integrate from 21 to 25.46 (which is 20.6 + 3 × 1.62), which is 3 standard deviations from our mean. Being three standard deviations from the mean will account for 99.7 percent of our total probability. The remaining 0.3 percent lies on either side of the distribution, so only half of that, 0.15 percent of our probability density, lies in the region greater than 25.46. So if we integrate from 21 to 25.46, we'll only be missing a tiny amount of probability in our result. Clearly, we could easily use R to integrate from 21 to something really safe such as 30, but this trick allows us to figure out what "really safe" means.

"N Sigma" Events

You may have heard an event being described in terms of *sigma events*, such as "the fall of the stock price was an eight-sigma event." What this expression means is that the observed data is eight standard deviations from the mean. We saw the progression of one, two, and three standard deviations from the mean in Table 12-1, which were values at 68, 95, and 99.7 percent, respectively. You can easily intuit from this that an eight-sigma event must be extremely unlikely. In fact, if you ever observe data that is five standard deviations from the mean, it's likely a good sign that your normal distribution is not modeling the underlying data accurately.

To show the growing rarity of an event as it increases by *n* sigma, say you are looking at events you might observe on a given day. Some are very common, such as waking up to the sunrise. Others are less common, such as waking up and it being your birthday. Table 12-2 shows how many days it would take to expect the event to happen per one sigma increase.

Table 12-2: Rarity of an Event as It Increases by *n* Sigma

(−/+) Distance from the mean	Expected every . . .
σ	3 days
2σ	3 weeks
3σ	1 year
4σ	4 decades
5σ	5 millennia
6σ	1.4 million years

So a three-sigma event is like waking up and realizing it's your birthday, but a six-sigma event is like waking up and realizing that a giant asteroid is crashing toward earth!

The Beta Distribution and the Normal Distribution

You may remember from Chapter 5 that the beta distribution allows us to estimate the true probability given that we have observed α desired outcomes and β undesired outcomes, where the total number of outcomes is $\alpha + \beta$. Based on that, you might take some issue with the notion that the normal distribution is truly the best method to model parameter estimation given that we know only the mean and standard deviation of any given data set. After all, we could describe a situation where $\alpha = 3$ and $\beta = 4$ by simply observing three values of 1 and four values of 0. This would give us $\mu = 0.43$ and $\sigma = 0.53$. We can then compare the beta distribution with $\alpha = 3$ and $\beta = 4$ to a normal distribution with $\mu = 0.43$ and $\sigma = 0.53$, as shown in Figure 12-9.

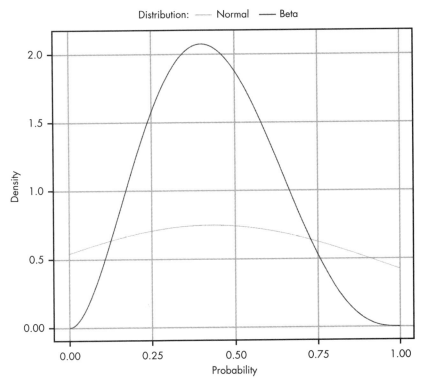

Figure 12-9: Comparing the beta distribution to the normal distribution

It's clear that these distributions are quite different. We can see that for both distributions the center of mass appears in roughly the same place, but the bounds for the normal distribution extend way beyond the limits of our graph. This demonstrates a key point: only when you know nothing about the data other than its mean and variance is it safe to assume a normal distribution.

For the beta distribution, we know that the value we're looking for must lie in the range 0 to 1. The normal distribution is defined from $-\infty$ to ∞, which often includes values that cannot possibly exist. However, in most cases this is not practically important because measurements out that far are essentially impossible in probabilistic terms. But for our example of measuring the probability of an event happening, this missing information is important for modeling our problem.

So, while the normal distribution is a very powerful tool, it is no substitute for having more information about a problem.

Wrapping Up

The normal distribution is an extension of using the mean for estimating a value from observations. The normal distribution combines the mean and the standard deviation to model how spread out our observations are from the mean. This is important because it allows us to reason about the error in our measurements in a probabilistic way. Not only can we use the mean to make our best guess, but we can also make probabilistic statements about ranges of possible values for our estimate.

Exercises

Try answering the following questions to see how well you understand the normal distribution. The solutions can be found at *https://nostarch.com/learnbayes/*.

1. What is the probability of observing a value five sigma greater than the mean or more?

2. A fever is any temperature greater than 100.4 degrees Fahrenheit. Given the following measurements, what is the probability that the patient has a fever?

   100.0, 99.8, 101.0, 100.5, 99.7

3. Suppose in Chapter 11 we tried to measure the depth of a well by timing coin drops and got the following values:

   2.5, 3, 3.5, 4, 2

   The distance an object falls can be calculated (in meters) with the following formula:

   $$distance = 1/2 \times G \times time^2$$

   where G is 9.8 m/s/s. What is the probability that the well is over 500 meters deep?

4. What is the probability there is no well (i.e., the well is really 0 meters deep)? You'll notice that probability is higher than you might expect, given your observation that there *is* a well. There are two good explanations for this probability being higher than it should. The first is that the normal distribution is a poor model for our measurements; the second is that, when making up numbers for an example, I chose values that you likely wouldn't see in real life. Which is more likely to you?

13

TOOLS OF PARAMETER ESTIMATION: THE PDF, CDF, AND QUANTILE FUNCTION

In this part so far, we've focused heavily on the building blocks of the normal distribution and its use in estimating parameters.

In this chapter, we'll dig in a bit more, exploring some mathematical tools we can use to make better claims about our parameter estimates. We'll walk through a real-world problem and see how to approach it in different ways using a variety of metrics, functions, and visualizations.

This chapter will cover more on the probability density function (PDF); introduce the cumulative distribution function (CDF), which helps us more easily determine the probability of ranges of values; and introduce quantiles, which divide our probability distributions into parts with equal probabilities. For example, a *percentile* is a 100-quantile, meaning it divides the probability distribution into 100 equal pieces.

Estimating the Conversion Rate for an Email Signup List

Say you run a blog and want to know the probability that a visitor to your blog will subscribe to your email list. In marketing terms, getting a user to perform a desired event is referred to as the *conversion event*, or simply a *conversion*, and the probability that a user will subscribe is the *conversion rate*.

As discussed in Chapter 5, we would use the beta distribution to estimate p, the probability of subscribing, when we know k, the number of people subscribed, and n, the total number of visitors. The two parameters needed for the beta distribution are α, which in this case represents the total subscribed (k), and β, representing the total not subscribed $(n - k)$.

When the beta distribution was introduced, you learned only the basics of what it looked like and how it behaved. Now you'll see how to use it as the foundation for parameter estimation. We want to not only make a single estimate for our conversion rate, but also come up with a range of possible values within which we can be very confident the real conversion rate lies.

The Probability Density Function

The first tool we'll use is the probability density function. We've seen the PDF several times so far in this book: in Chapter 5 where we talked about the beta distribution; in Chapter 9 when we used PDFs to combine Bayesian priors; and once again in Chapter 12, when we talked about the normal distribution. The PDF is a function that takes a value and returns the probability of that value.

In the case of estimating the true conversion rate for your email list, let's say for the first 40,000 visitors, you get 300 subscribers. The PDF for our problem is the beta distribution where $\alpha = 300$ and $\beta = 39,700$:

$$\mathrm{Beta}\left(x; 300, 39700\right) = \frac{x^{300-1}\left(1 - x\right)^{39700-1}}{\mathrm{beta}\left(300, 39700\right)}$$

We've spent a lot of time talking about using the mean as a good estimate for a measurement, given some uncertainty. Most PDFs have a mean, which we compute specifically for the beta distribution as follows:

$$\mu_{\mathrm{Beta}} = \frac{\alpha}{\alpha + \beta}$$

This formula is relatively intuitive: simply divide the number of outcomes we care about (300) by the total number of outcomes (40,000). This is the same mean you'd get if you simply considered each email an observation of 1 and all the others an observation of 0 and then averaged them out.

The mean is our first stab at estimating a parameter for the true conversion rate. But we'd still like to know other possible values for our conversion rate. Let's continue exploring the PDF to see what else we can learn.

Visualizing and Interpreting the PDF

The PDF is usually the go-to function for understanding a distribution of probabilities. Figure 13-1 illustrates the PDF for the blog conversion rate's beta distribution.

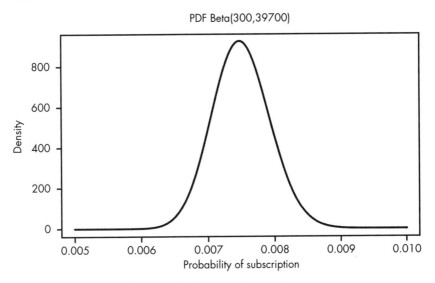

Figure 13-1: Visualizing the beta PDF for our beliefs in the true conversion rate

What does this PDF represent? From the data we know that the blog's average conversion rate is simply

$$\frac{\text{subscribed}}{\text{visited}} = \frac{300}{40,000} = 0.0075$$

or the *mean* of our distribution. It seems unlikely that the conversion rate is *exactly* 0.0075 rather than, say, 0.00751. We know the total area under the curve of the PDF must add up to 1, since this PDF represents the probability of all possible estimates. We can estimate ranges of values for our true conversion rate by looking at the area under the curve for the ranges we care about. In calculus, this area under the curve is the *integral*, and it tells us how much of the total probability is in the region of the PDF we're interested in. This is exactly like how we used integration with the normal distribution in the prior chapter.

Given that we have uncertainty in our measurement, and we have a mean, it could be useful to investigate how much more likely it is that the true conversion rate is 0.001 higher or lower than the mean of 0.0075 we observed. Doing so would give us an acceptable margin of error (that is, we'd be happy with any values in this range). To do this, we can calculate the probability of the actual rate being lower than 0.0065, and the probability of the actual rate being higher than 0.0085, and then compare them. The probability that our conversion rate is actually much lower than our observations is calculated like so:

$$P(\text{much lower}) = \int_0^{0.0065} \text{Beta}(300, 39700) = 0.008$$

Remember that when we take the integral of a function, we are just summing all the little pieces of our function. So, if we take the integral from 0 to 0.0065 for the beta distribution with an α of 300 and a β of 39,700, we are adding up all the probabilities for the values in this range and determining the probability that our true conversion rate is somewhere between 0 and 0.0065.

We can ask questions about the other extreme as well, such as: how likely is it that we actually got an unusually bad sample and our true conversion rate is much higher, such as a value greater than, say, 0.0085 (meaning a better conversion rate than we had hoped)?

$$P(\text{much higher}) = \int_{0.0085}^{1} \text{Beta}(300, 397000) = 0.012$$

Here we are integrating from 0.0085 to the largest possible value, which is 1, to determine the probability that our true value lies somewhere in this range. So, in this example, the probability that our conversion rate is 0.001 higher or more than we observed is actually more likely than the probability that it is 0.001 less or worse than observed. This means that if we had to make a decision with the limited data we have, we could still calculate how much likelier one extreme is than the other:

$$\frac{P(\text{much higher})}{P(\text{much lower})} = \frac{\int_{0.0085}^{1} \text{Beta}(300, 397000)}{\int_0^{0.0065} \text{Beta}(300, 39700)} = \frac{0.012}{0.008} = 1.5$$

Thus, it's 50 percent more likely that our true conversion rate is greater than 0.0085 than that it's lower than 0.0065.

Working with the PDF in R

In this book we've already used two R functions for working with PDFs, dnorm() and dbeta(). For most well-known probability distributions, R supports an equivalent dfunction() function for calculating the PDF.

Functions like dbeta() are also useful for approximating the continuous PDF—for example, when you want to quickly plot out values like these:

```
xs <- seq(0.005,0.01,by=0.00001)
xs.all <- seq(0,1,by=0.0001)
plot(xs,dbeta(xs,300,40000-300),type='l',lwd=3,
     ylab="density",
     xlab="probability of subscription",
     main="PDF Beta(300,39700)")
```

NOTE   *To understand the plotting code, see Appendix A.*

In this example code, we're creating a sequence of values that are each 0.00001 apart—small, but not infinitely small, as they would be in a truly continuous distribution. Nonetheless, when we plot these values, we see something that looks close enough to a truly continuous distribution (as shown earlier in Figure 13-1).

Introducing the Cumulative Distribution Function

The most common mathematical use of the PDF is in integration, to solve for probabilities associated with various ranges, just as we did in the previous section. However, we can save ourselves a lot of effort with the *cumulative distribution function (CDF)*, which sums all parts of our distribution, replacing a lot of calculus work.

The CDF takes in a value and returns the probability of getting that value or lower. For example, the CDF for Beta(300,397000) when $x = 0.0065$ is approximately 0.008. This means that the probability of the true conversion rate being 0.0065 or less is 0.008.

The CDF gets this probability by taking the cumulative area under the curve for the PDF (for those comfortable with calculus, the CDF is the *antiderivative* of the PDF). We can summarize this process in two steps: (1) figure out the cumulative area under the curve for each value of the PDF, and (2) plot those values. That's our CDF. The value of the curve at any given x-value is the probability of getting a value of x or lower. At 0.0065, the value of the curve would be 0.008, just as we calculated earlier.

To understand how this works, let's break the PDF for our problem into chunks of 0.0005 and focus on the region of our PDF that has the most probability density: the region between 0.006 and 0.009.

Figure 13-2 shows the cumulative area under the curve for the PDF of Beta(300,39700). As you can see, our cumulative area under the curve takes into account all of the area in the pieces to its left.

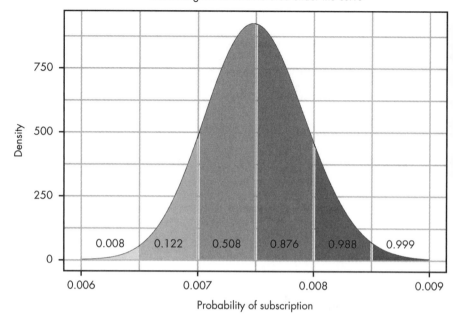

Figure 13-2: Visualizing the cumulative area under the curve

Mathematically speaking, Figure 13-2 represents the following sequence of integrals:

$$\int_0^{0.0065} \text{Beta}(300, 397000)$$

$$\int_0^{0.0065} \text{Beta}(300, 397000) + \int_{0.0065}^{0.007} \text{Beta}(300, 397000)$$

$$\int_0^{0.0065} \text{Beta}(300, 397000) + \int_{0.0065}^{0.007} \text{Beta}(300, 397000) + \int_{0.007}^{0.0075} \text{Beta}(300, 397000)$$

(And so on)

Using this approach, as we move along the PDF, we take into account an increasingly higher probability until our total area is 1, or complete certainty. To turn this into the CDF, we can imagine a function that looks at only these areas under the curve. Figure 13-3 shows what happens if we plot the area under the curve for each of our points, which are 0.0005 apart.

Now we have a way of visualizing just how the cumulative area under the curve changes as we move along the values for our PDF. Of course, the problem is that we're using these discrete chunks. In reality, the CDF just uses infinitely small pieces of the PDF, so we get a nice smooth line (see Figure 13-4).

In our example, we derived the CDF visually and intuitively. Deriving the CDF mathematically is much more difficult, and often leads to very complicated equations. Luckily, we typically use code to work with the CDF, as we'll see in a few more sections.

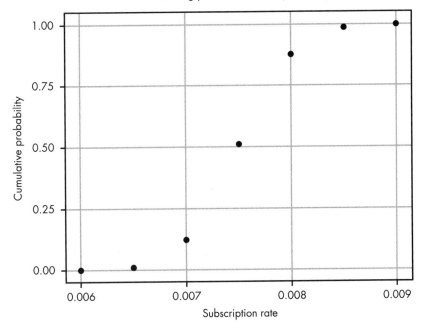

Figure 13-3: Plotting just the cumulative probability from Figure 13-2

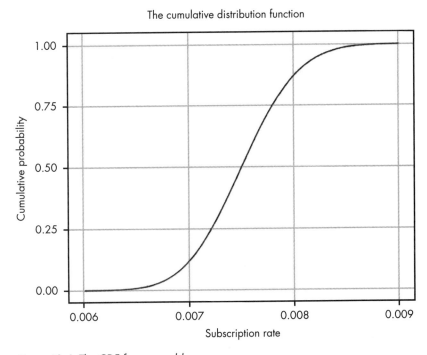

Figure 13-4: The CDF for our problem

Visualizing and Interpreting the CDF

The PDF is most useful visually for quickly estimating where the peak of a distribution is, and for getting a rough sense of the width (variance) and shape of a distribution. However, with the PDF it is very difficult to reason about the probability of various ranges visually. The CDF is a much better tool for this. For example, we can use the CDF in Figure 13-4 to visually reason about a much wider range of probabilistic estimates for our problem than we can using the PDF alone. Let's go over a few visual examples of how we can use this amazing mathematical tool.

Finding the Median

The median is the point in the data at which half the values fall on one side and half on the other—it is the exact middle *value* of our data. In other words, the probability of a value being greater than the median and the probability of it being less than the median are both 0.5. The median is particularly useful for summarizing the data in cases where it contains extreme values.

Unlike the mean, computing the median can actually be pretty tricky. For small, discrete cases, it's as simple as putting your observations in order and selecting the value in the middle. But for continuous distributions like our beta distribution, it's a little more complicated.

Thankfully, we can easily spot the median on a visualization of the CDF. We can simply draw a line from the point where the cumulative probability is 0.5, meaning 50 percent of the values are below this point and 50 percent are above. As Figure 3-5 illustrates, the point where this line intersects the x-axis gives us our median!

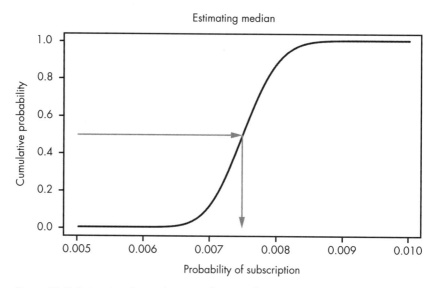

Figure 13-5: Estimating the median visually using the CDF

We can see that the median for our data is somewhere between 0.007 and 0.008 (this happens to be very close the mean of 0.0075, meaning the data isn't particularly skewed).

Approximating Integrals Visually

When working with ranges of probabilities, we'll often want to know the probability that the true value lies somewhere between some value y and some value x.

We can solve this kind of problem using integration, but even if R makes solving integrals easier, it's very time-consuming to make sense of the data and to constantly rely on R to compute integrals. Since all we want is a rough estimate that the probability of a visitor subscribing to the blog falls within a particular range, we don't need to use integration. The CDF makes it very easy to eyeball whether or not a certain range of values has a very high probability or a very low probability of occurring.

To estimate the probability that the conversion rate is between 0.0075 and 0.0085, we can trace lines from the x-axis at these points, then see where they meet up with the y-axis. The distance between the two points is the approximate integral, as shown in Figure 13-6.

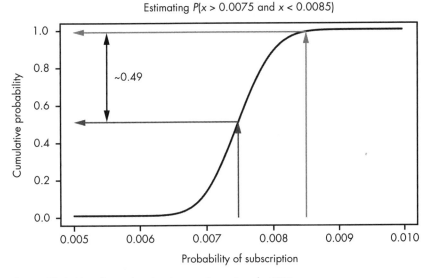

Estimating $P(x > 0.0075$ and $x < 0.0085)$

Figure 13-6: Visually performing integration using the CDF

We can see that on the y-axis these values range from roughly 0.5 to 0.99, meaning that there is approximately a 49 percent chance that our true conversion rate lies somewhere between these two values. The best part is we didn't have to do any integration! This is, of course, because the CDF represents the integral from the minimum of our function to all possible values.

So, since nearly all of the probabilistic questions about a parameter estimate involve knowing the probability associated with certain ranges of beliefs, the CDF is often a far more useful visual tool than the PDF.

Estimating Confidence Intervals

Looking at the probability of ranges of values leads us to a very important concept in probability: the *confidence interval*. A confidence interval is a lower and upper bound of values, typically centered on the mean, describing a range of high probability, usually 95, 99, or 99.9 percent. When we say something like "The 95 percent confidence interval is from 12 to 20," what we mean is that there is a 95 percent probability that our true measurement is somewhere between 12 and 20. Confidence intervals provide a good method of describing the range of possibilities when we're dealing with uncertain information.

NOTE   *In Bayesian statistics what we are calling a "confidence interval" can go by a few other names, such as "critical region" or "critical interval." In some more traditional schools of statistics, "confidence interval" has a slightly different meaning, which is beyond the scope of this book.*

We can estimate confidence intervals using the CDF. Say we wanted to know the range that covers 80 percent of the possible values for the true conversion rate. We solve this problem by combining our previous approaches: we draw lines at the y-axis from 0.1 and 0.9 to cover 80 percent, and then simply see where on the x-axis these intersect with our CDF, as shown in Figure 13-7.

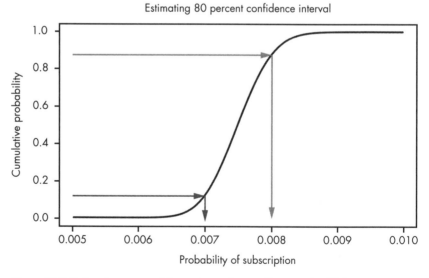

Figure 13-7: Estimating our confidence intervals visually using the CDF

As you can see, the x-axis is intersected at roughly 0.007 and 0.008, which means that there's an 80 percent chance that our true conversion rate falls somewhere between these two values.

Using the CDF in R

Just as nearly all major PDFs have a function starting with d, like dnorm(), CDF functions start with p, such as pnorm(). In R, to calculate the probability that Beta(300,39700) is less than 0.0065, we can simply call pbeta() like this:

```
pbeta(0.0065,300,39700)
> 0.007978686
```

And to calculate the true probability that the conversion rate is greater than 0.0085, we can do the following:

```
pbeta(1,300,39700) - pbeta(0.0085,300,39700)
> 0.01248151
```

The great thing about CDFs is that it doesn't matter if your distribution is discrete or continuous. If we wanted to determine the probability of getting three or fewer heads in five coin tosses, for example, we would use the CDF for the binomial distribution like this:

```
pbinom(3,5,0.5)
> 0.8125
```

The Quantile Function

You might have noticed that the median and confidence intervals we took visually with the CDF are not easy to do mathematically. With the visualizations, we simply drew lines from the y-axis and used those to find a point on the x-axis.

Mathematically, the CDF is like any other function in that it takes an x value, often representing the value we're trying to estimate, and gives us a y value, which represents the cumulative probability. But there is no obvious way to do this in reverse; that is, we can't give the same function a y to get an x. As an example, imagine we have a function that squares values. We know that square(3) = 9, but we need an entirely new function—the square root function—to know that the square root of 9 is 3.

However, reversing the function is *exactly what we did* in the previous section to estimate the median: we looked at the y-axis for 0.5, then traced it back to the x-axis. What we've done visually is compute the *inverse* of the CDF.

While computing the inverse of the CDF visually is easy for estimates, we need a separate mathematical function to compute it for exact values. The inverse of the CDF is an incredibly common and useful tool called the *quantile function*. To compute an exact value for our median and confidence interval, we need to use the quantile function for the beta distribution. Just like the CDF, the quantile function is often very tricky to derive and use mathematically, so instead we rely on software to do the hard work for us.

Visualizing and Understanding the Quantile Function

Because the quantile function is simply the inverse of the CDF, it just looks like the CDF rotated 90 degrees, as shown in Figure 13-8.

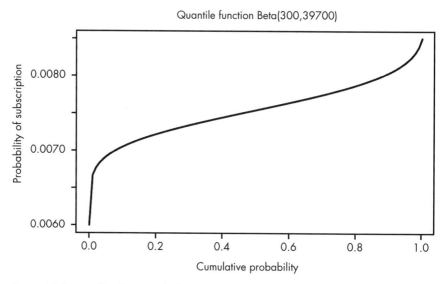

Figure 13-8: Visually, the quantile function is just a rotation of the CDF.

Whenever you hear phrases like:

"The top 10 percent of students . . . "

"The bottom 20 percent of earners earn less than . . . "

"The top quartile has notably better performance than . . . "

you're talking about values that are found using the quantile function. To look up a quantile visually, just find the quantity you're interested in on the x-axis and see where it meets the y-axis. The value on the y-axis is the value for that quantile. Keep in mind that if you're talking about the "top 10 percent," you really want the 0.9 quantile.

Calculating Quantiles in R

R also includes the function qnorm() for calculating quantiles. This function is very useful for quickly answering questions about what values are bounds of our probability distribution. For example, if we want to know the value that 99.9 percent of the distribution is less than, we can use qbeta() with the quantile we're interested in calculating as the first argument, and the alpha and beta parameters of our beta distribution as the second and third arguments, like so:

```
qbeta(0.999,300,39700)
> 0.008903462
```

The result is 0.0089, meaning we can be 99.9 percent certain that the true conversion rate for our emails is less than 0.0089. We can then use the quantile function to quickly calculate exact values for confidence intervals for our estimates. To find the 95 percent confidence interval, we can find the values greater than the 2.5 percent lower quantile and the values lower than the 97.5 percent upper quantile, and the interval between them is the 95 percent confidence interval (the unaccounted region totals 5 percent of the probability density at both extremes). We can easily calculate these for our data with qbeta():

Our lower bound is qbeta(0.025,300,39700) = 0.0066781

Our upper bound is qbeta(0.975,300,39700) = 0.0083686

Now we can confidently say that we are 95 percent certain that the real conversion rate for blog visitors is somewhere between 0.67 percent and 0.84 percent.

We can, of course, increase or decrease these thresholds depending on how certain we want to be. Now that we have all of the tools of parameter estimation, we can easily pin down an exact range for the conversion rate. The great news is that we can also use this to predict ranges of values for future events.

Suppose an article on your blog goes viral and gets 100,000 visitors. Based on our calculations, we know that we should expect between 670 and 840 new email subscribers.

Wrapping Up

We've covered a lot of ground and touched on the interesting relationship between the probability density function (PDF), cumulative distribution function (CDF), and the quantile function. These tools form the basis of how we can estimate parameters and calculate our confidence in those estimations. That means we can not only make a good guess as to what an unknown value might be, but also determine confidence intervals that very strongly represent the possible values for a parameter.

Exercises

Try answering the following questions to see how well you understand the tools of parameter estimation. The solutions can be found at *https://nostarch.com/learnbayes/*.

1. Using the code example for plotting the PDF on page 127, plot the CDF and quantile functions.

2. Returning to the task of measuring snowfall from Chapter 10, say you have the following measurements (in inches) of snowfall:

   7.8, 9.4, 10.0, 7.9, 9.4, 7.0, 7.0, 7.1, 8.9, 7.4

   What is your 99.9 percent confidence interval for the true value of snowfall?

3. A child is going door to door selling candy bars. So far she has visited 30 houses and sold 10 candy bars. She will visit 40 more houses today. What is the 95 percent confidence interval for how many candy bars she will sell the rest of the day?

14

PARAMETER ESTIMATION WITH PRIOR PROBABILITIES

In the previous chapter, we looked at using some important mathematical tools to estimate the conversion rate for blog visitors subscribing to an email list. However, we haven't yet covered one of the most important parts of parameter estimation: using our existing beliefs about a problem.

In this chapter, you'll see how we can use our prior probabilities, combined with observed data, to come up with a better estimate that blends existing knowledge with the data we've collected.

Predicting Email Conversion Rates

To understand how the beta distribution changes as we gain information, let's look at another conversion rate. In this example, we'll try to figure out the rate at which your subscribers click a given link once they've opened an email from you. Most companies that provide email list management services tell you, in real time, how many people have opened an email and clicked the link.

Our data so far tells us that of the first five people that open an email, two of them click the link. Figure 14-1 shows our beta distribution for this data.

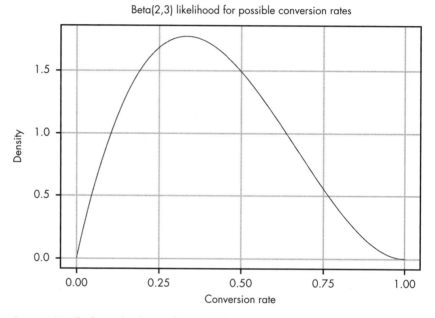

Figure 14-1: The beta distribution for our observations so far

Figure 14-1 shows Beta(2,3). We used these numbers because two people clicked and three did not click. Unlike in the previous chapter, where we had a pretty narrow spike in possible values, here we have a huge range of possible values for the true conversion rate because we have very little information to work with. Figure 14-2 shows the CDF for this data, to help us more easily reason about these probabilities.

The 95 percent confidence interval (i.e., a 95 percent chance that our true conversion rate is somewhere in that range) is marked to make it easier to see. At this point our data tells us that the true conversion rate could be anything between 0.05 and 0.8! This is a reflection of how little information we've actually acquired so far. Given that we've had two conversions, we know the true rate can't be 0, and since we've had three non-conversions, we also know it can't be 1. Almost everything else is fair game.

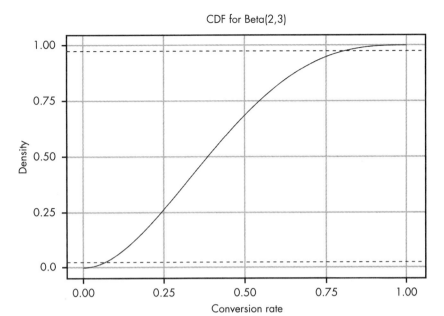

Figure 14-2: CDF for our observation

Taking in Wider Context with Priors

But wait a second—you may be new to email lists, but an 80 percent click-through rate sounds pretty unlikely. I subscribe to plenty of lists, but I definitely don't click through to the content 80 percent of the time that I open the email. Taking that 80 percent rate at face value seems naive when I consider my own behavior.

As it turns out, your email service provider thinks it's suspicious too. Let's look at some wider context. For blogs listed in the same category as yours, the provider's data claims that on average only 2.4 percent of people who open emails click through to the content.

In Chapter 9, you learned how we could use past information to modify our belief that Han Solo can successfully navigate an asteroid field. Our data tells us one thing, but our background information tells us another. As you know by now, in Bayesian terms the data we have observed is our *likelihood,* and the external context information—in this case from our personal experience and our email service—is our *prior probability.* Our challenge now is to figure out how to model our prior. Luckily, unlike the case with Han Solo, we actually have some data here to help us.

The conversion rate of 2.4 percent from your email provider gives us a starting point: now we know we want a beta distribution whose mean is roughly 0.024. (The mean of a beta distribution is $\alpha / (\alpha + \beta)$.) However, this still leaves us with a range of possible options: Beta(1,41), Beta(2,80), Beta(5,200), Beta(24,976), and so on. So which should we use? Let's plot some of these out and see what they look like (Figure 14-3).

Possible priors for email conversion rates

Distribution: ——— Beta(1,41) ······ Beta(2,80) – – – Beta(5,200)

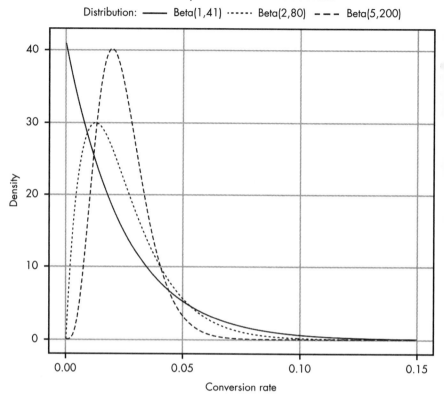

Figure 14-3: Comparing different possible prior probabilities

As you can see, the lower the combined $\alpha + \beta$, the wider our distribution. The problem now is that even the most liberal option we have, Beta(1,41), seems a little too pessimistic, as it puts a lot of our probability density in very low values. We'll stick with this distribution nonetheless, since it is based on the 2.4 percent conversion rate in the data from the email provider, and is the weakest of our priors. Being a "weak" prior means it will be more easily overridden by actual data as we collect more of it. A stronger prior, like Beta(5,200), would take more evidence to change (we'll see how this happens next). Deciding whether or not to use a strong prior is a judgment call based on how well you expect the prior data to describe what you're currently doing. As we'll see, even a weak prior can help keep our estimates more realistic when we're working with small amounts of data.

Remember that, when working with the beta distribution, we can calculate our posterior distribution (the combination of our likelihood and our prior) by simply adding together the parameters for the two beta distributions:

$$\text{Beta}\left(\alpha_{\text{posterior}}, \beta_{\text{posterior}}\right) = \text{Beta}\left(\alpha_{\text{likelihood}} + \alpha_{\text{prior}}, \beta_{\text{likelihood}} + \beta_{\text{prior}}\right)$$

Using this formula, we can compare our beliefs with and without priors, as shown in Figure 14-4.

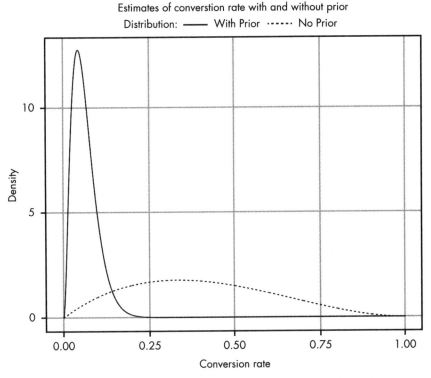

Figure 14-4: Comparing our likelihood (no prior) to our posterior (with prior)

Wow! That's quite sobering. Even though we're working with a relatively weak prior, we can see that it has made a huge impact on what we believe are realistic conversion rates. Notice that for the likelihood with no prior, we have some belief that our conversion rate could be as high as 80 percent. As mentioned, this is highly suspicious; any experienced email marketer would tell you than an 80 percent conversion rate is unheard of. Adding a prior to our likelihood adjusts our beliefs so that they become much more reasonable. But I still think our updated beliefs are a bit pessimistic. Maybe the email's true conversion rate isn't 40 percent, but it still might be better than this current posterior distribution suggests.

How can we prove that our blog has a better conversion rate than the sites in the email provider's data, which have a 2.4 percent conversion rate? The way any rational person does: with more data! We wait a few hours to gather more results and now find that out of 100 people who opened your email, 25 have clicked the link! Let's look at the difference between our new posterior and likelihood, shown in Figure 14-5.

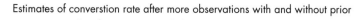

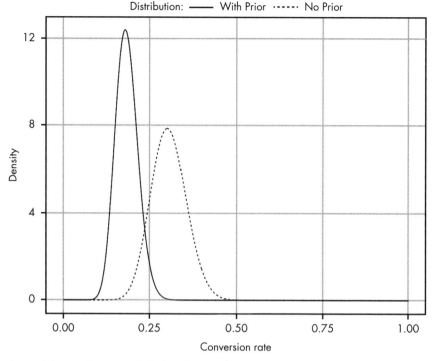

Figure 14-5: Updating our beliefs with more data

As we continue to collect data, we see that our posterior distribution using a prior is starting to shift toward the one without the prior. Our prior is still keeping our ego in check, giving us a more conservative estimate for the true conversion rate. However, as we add evidence to our likelihood, it starts to have a bigger impact on what our posterior beliefs look like. In other words, the additional observed data is doing what it should: slowly swaying our beliefs to align with what it suggests. So let's wait overnight and come back with even more data!

In the morning we find that 300 subscribers have opened their email, and 86 of those have clicked through. Figure 14-6 shows our updated beliefs.

What we're witnessing here is the most important point about Bayesian statistics: the more data we gather, the more our prior beliefs become diminished by evidence. When we had almost no evidence, our likelihood proposed some rates we know are absurd (e.g., 80 percent click-through), both intuitively and from personal experience. In light of little evidence, our prior beliefs squashed any data we had.

But as we continue to gather data that disagrees with our prior, our posterior beliefs shift toward what our own collected data tells us and away from our original prior.

Another important takeaway is that we started with a pretty weak prior. Even then, after just a day of collecting a relatively small set of information, we were able to find a posterior that seems much, much more reasonable.

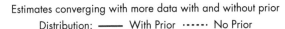

Estimates converging with more data with and without prior
Distribution: —— With Prior ······ No Prior

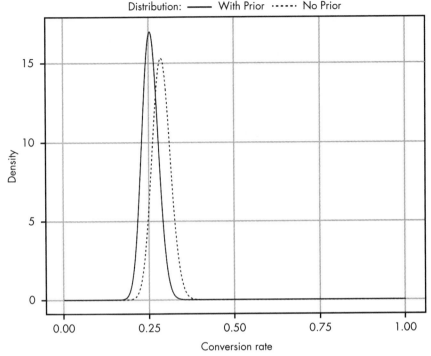

Figure 14-6: Our posterior beliefs with even more data added

The prior probability distribution in this case helped tremendously with keeping our estimate much more realistic in the absence of data. This prior probability distribution was based on real data, so we could be fairly confident that it would help us get our estimate closer to reality. However, in many cases we simply don't have any data to back up our prior. So what do we do then?

Prior as a Means of Quantifying Experience

Because we knew the idea of an 80 percent click-through rate for emails was laughable, we used data from our email provider to come up with a better estimate for our prior. However, even if we didn't have data to help establish our prior, we could still ask someone with a marketing background to help us make a good estimate. A marketer might know from personal experience that you should expect about a 20 percent conversion rate, for example.

Given this information from an experienced professional, you might choose a relatively weak prior like Beta(2,8) to suggest that the expected conversion rate should be around 20 percent. This distribution is just a guess, but the important thing is that we can quantify this assumption. For nearly every business, experts can often provide powerful prior information based simply on previous experience and observation, even if they have no training in probability specifically.

By quantifying this experience, we can get more accurate estimates and see how they can change from expert to expert. For example, if a marketer is certain that the true conversion rate should be 20 percent, we might model this belief as Beta(200,800). As we gather data, we can compare models and create multiple confidence intervals that quantitatively model any expert beliefs. Additionally, as we gain more and more information, the difference due to these prior beliefs will decrease.

Is There a Fair Prior to Use When We Know Nothing?

There are certain schools of statistics that teach that you should always add 1 to both α and β when estimating parameters with no other prior. This corresponds to using a very weak prior that holds that each outcome is equally likely: Beta(1,1). The argument is that this is the "fairest" (i.e., weakest) prior we can come up with in the absence of information. The technical term for a fair prior is a *noninformative prior*. Beta(1,1) is illustrated in Figure 14-7.

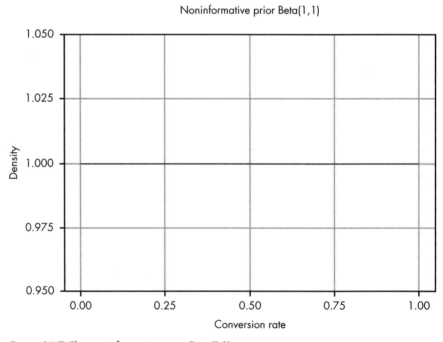

Figure 14-7: The noninformative prior Beta(1,1)

As you can see, this is a perfectly straight line, so that all outcomes are then equally likely and the mean likelihood is 0.5. The idea of using a noninformative prior is that we can add a prior to help smooth out our estimate, but that prior isn't biased toward any particular outcome. However, while this may initially seem like the fairest way to approach the problem, even this very weak prior can lead to some strange results when we test it out.

Take, for example, the probability that the sun will rise tomorrow. Say you are 30 years old, and so you've experienced about 11,000 sunrises in your lifetime. Now suppose someone asks the probability that the sun will rise tomorrow. You want to be fair and use a noninformative prior, Beta(1,1). The distribution that represents your belief that the sun will *not* rise tomorrow would be Beta(1,11001), based on your experiences. While this gives a very low probability for the sun not rising tomorrow, it also suggests that we would expect to see the sun *not* rise at least once by the time you reach 60 years old. The so-called "noninformative" prior is providing a pretty strong opinion about how the world works!

You could argue that this is only a problem because we understand celestial mechanics, so we already have strong prior information we can't forget. But the real problem is that *we've never observed the case where the sun doesn't rise.* If we go back to our likelihood function without the noninformative prior, we get Beta(0,11000).

However, when either α or $\beta \leq 0$, the beta distribution is *undefined*, which means that the correct answer to "What is the probability that the sun will rise tomorrow?" is that the question doesn't make sense because we've never seen a counterexample.

As another example, suppose you found a portal that transported you and a friend to a new world. An alien creature appears before you and fires a strange-looking gun at you that just misses. Your friend asks you, "What's the probability that the gun will misfire?" This is a completely alien world and the gun looks strange and organic, so you know nothing about its mechanics at all.

This is, in theory, the ideal scenario for using a noninformative prior, since you have absolutely no prior information about this world. If you add your noninformative prior, you get a posterior Beta(1,2) probability that the gun will misfire (we observed $\alpha = 0$ misfires and $\beta = 1$ successful fires). This distribution tells us the mean posterior probability of a misfire is 1/3, which seems astoundingly high given that you don't even know *if* the strange gun can misfire. Again, even though Beta(0,1) is undefined, using it seems like the rational approach to this problem. In the absence of sufficient data and any prior information, your only honest option is to throw your hands in the air and tell your friend, "I have no clue how to even reason about that question!"

The best priors are backed by data, and there is never really a true "fair" prior when you have a total lack of data. Everyone brings to a problem their own experiences and perspective on the world. The value of Bayesian reasoning, even when you are subjectively assigning priors, is that you are quantifying your subjective belief. As we'll see later in the book, this means you can compare your prior to other people's and see how well it explains the world around you. A Beta(1,1) prior is sometimes used in practice, but you should use it only when you earnestly believe that the two possible outcomes are, as far as you know, equally likely. Likewise, no amount of mathematics can make up for absolute ignorance. If you have no data and no prior understanding of a problem, the only honest answer is to say that you can't conclude anything at all until you know more.

All that said, it's worth noting that this topic of whether to use Beta(1,1) or Beta(0,0) has a long history, with many great minds arguing various positions. Thomas Bayes (namesake of Bayes' theorem) hesitantly believed in Beta(1,1); the great mathematician Simon-Pierre Laplace was quite certain Beta(1,1) was correct; and the famous economist John Maynard Keynes thought using Beta(1,1) was so preposterous that it discredited all of Bayesian statistics!

Wrapping Up

In this chapter, you learned how to incorporate prior information about a problem to arrive at much more accurate estimates for unknown parameters. When we have only a little information about a problem, we can easily get probabilistic estimates that seem impossible. But we might have prior information that can help us make better inferences from that small amount of data. By adding this information to our estimates, we get much more realistic results.

Whenever possible, it's best to use a prior probability distribution based on actual data. However, often we won't have data to support our problem, but we either have personal experience or can turn to experts who do. In these cases, it's perfectly fine to estimate a probability distribution that corresponds to your intuition. Even if you're wrong, you'll be wrong in a way that is recorded quantitatively. Most important, even if your prior is wrong, it will eventually be overruled by data as you collect more observations.

Exercises

Try answering the following questions to see how well you understand priors. The solutions can be found at *https://nostarch.com/learnbayes/*.

1. Suppose you're playing air hockey with some friends and flip a coin to see who starts with the puck. After playing 12 times, you realize that the friend who brings the coin almost always seems to go first: 9 out of 12 times. Some of your other friends start to get suspicious. Define prior probability distributions for the following beliefs:

   • One person who weakly believes that the friend is cheating and the true rate of coming up heads is closer to 70 percent.

   • One person who very strongly trusts that the coin is fair and provided a 50 percent chance of coming up heads.

   • One person who strongly believes the coin is biased to come up heads 70 percent of the time.

2. To test the coin, you flip it 20 more times and get 9 heads and 11 tails. Using the priors you calculated in the previous question, what are the updated posterior beliefs in the true rate of flipping a heads in terms of the 95 percent confidence interval?

PART IV

HYPOTHESIS TESTING: THE HEART OF STATISTICS

15

FROM PARAMETER ESTIMATION TO HYPOTHESIS TESTING: BUILDING A BAYESIAN A/B TEST

In this chapter, we're going to build our first hypothesis test, an *A/B test*. Companies often use A/B tests to try out product web pages, emails, and other marketing materials to determine which will work best for customers. In this chapter, we'll test our belief that removing an image from an email will increase the *click-through rate* against the belief that removing it will hurt the click-through rate.

Since we already know how to estimate a single unknown parameter, all we need to do for our test is estimate both parameters—that is, the conversion rates of each email. Then we'll use R to run a Monte Carlo simulation and determine which hypothesis is likely to perform better—in other words, which variant, A or B, is superior. A/B tests can be performed using classical statistical techniques such as *t*-tests, but building our test the Bayesian way will help us understand each part of it intuitively and give us more useful results as well.

We've covered the basics of parameter estimation pretty well at this point. We've seen how to use the PDF, CDF, and quantile functions to learn the likelihood of certain values, and we've seen how to add a Bayesian prior to our estimate. Now we want to use our estimates to compare *two* unknown parameters.

Setting Up a Bayesian A/B Test

Keeping with our email example from the previous chapter, imagine we want to see whether adding an image helps or hurts the conversion rate for our blog. Previously, the weekly email has included some image. For our test we're going to send one variant with images like usual, and another without images. The test is called an *A/B test* because we are comparing variant A (with image) and variant B (without) to determine which one performs better.

Let's assume at this point we have 600 blog subscribers. Because we want to exploit the knowledge gained during this experiment, we're only going to be running our test on 300 of them; that way, we can send the remaining 300 subscribers what we believe to be the most effective variant of the email.

The 300 people we're going to test will be split up into two groups, A and B. Group A will receive the usual email with a big picture at the top, and group B will receive an email with no picture. The hope is that a simpler email will feel less "spammy" and encourage users to click through to the content.

Finding Our Prior Probability

Next, we need to figure out what prior probability we're going to use. We've run an email campaign every week, so from that data we have a reasonable expectation that the probability of clicking the link to the blog on any given email should be around 30 percent. To make things simple, we'll use the same prior for both variants. We'll also choose a pretty weak version of our prior distribution, meaning that it considers a wider range of conversion rates to be probable. We're using a weak prior because we don't really know how well we expect B to do, and this is a new email campaign, so other factors could cause a better or worse conversion. We'll settle on Beta(3,7) for our prior probability distribution. This distribution allows us to represent a beta distribution where 0.3 is the mean, but a wide range of possible alternative rates are considered. We can see this distribution in Figure 15-1.

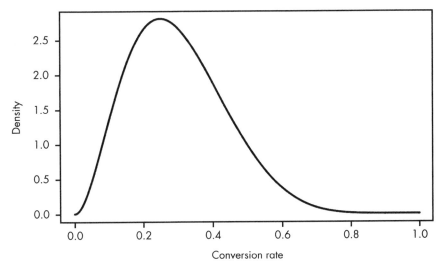

Figure 15-1: Visualizing our prior probability distribution

All we need now is our likelihood, which means we need to collect data.

Collecting Data

We send out our emails and get the results in Table 15-1.

Table 15-1: Email Click-through Rates

	Clicked	Not clicked	Observed conversion rate
Variant A	36	114	0.24
Variant B	50	100	0.33

We can treat each of these variants as a separate parameter we're trying to estimate. In order to arrive at a posterior distribution for each, we need to combine both their likelihood distribution and prior distribution. We've already decided that the prior for these distributions should be Beta(3,7), representing a relatively weak belief in what possible values we expect the conversion rate to be, given no additional information. We say this is a weak belief because we don't believe very strongly in a particular range of values, and consider all possible rates with a reasonably high probability. For the likelihood of each, we'll again use the beta distribution, making α the number of times the link was clicked through and β the number of times it was not.

Recall that:

$$\mathrm{Beta}\left(\alpha_{\mathrm{posterior}}, \beta_{\mathrm{posterior}}\right) = \mathrm{Beta}\left(\alpha_{\mathrm{prior}} + \alpha_{\mathrm{likelihood}}, \beta_{\mathrm{prior}} + \beta_{\mathrm{likelihood}}\right)$$

Variant A will be represented by Beta(36+3,114+7) and variant B by Beta(50+3,100+7). Figure 15-2 shows the estimates for each parameter side by side.

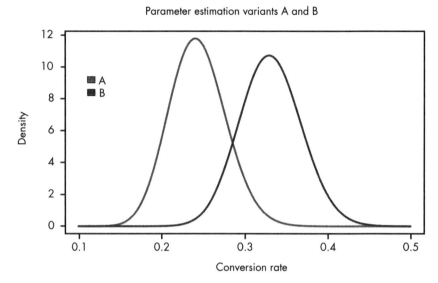

Figure 15-2: Beta distributions for our estimates for both variants of our email

Clearly, our data suggests that variant B is superior, in that it garners a higher conversion rate. However, from our earlier discussion on parameter estimation, we know that the true conversion rate is one of a range of possible values. We can also see here that there's an overlap between the possible true conversion rates for A and B. What if we were just unlucky in our A responses, and A's true conversion rate is in fact much higher? What if we were also just lucky with B, and its conversion rate is in fact much lower? It's easy to see a possible world in which A is actually the better variant, even though it did worse on our test. So the real question is: how sure can we be that B is the better variant? This is where the Monte Carlo simulation comes in.

Monte Carlo Simulations

The accurate answer to which email variant generates a higher click-through rate lies somewhere in the intersection of the distributions of A and B. Fortunately, we have a way to figure it out: a Monte Carlo simulation. A *Monte Carlo simulation* is any technique that makes use of random sampling to solve a problem. In this case, we're going to randomly sample from the two distributions, where each sample is chosen based on its probability in the distribution so that samples in a high-probability region will appear more frequently. For example, as we can see in Figure 15-2, a value *greater* than 0.2 is far more likely to be sampled from A than a value less than 0.2. However, a random sample from distribution B is nearly certain to be above 0.2. In our random sampling, we might pick out a value of 0.2 for variant A and 0.35 for variant B. Each sample is random, and based on the relative probability of values in the A and B distributions. The values 0.2 for A and 0.35 for B both could be the true conversion rate for each variant based on the evidence we've observed. This individual sampling from the two distributions confirms the belief that variant B is, in fact, superior to A, since 0.35 is larger than 0.2.

However, we could also sample 0.3 for variant A and 0.27 for variant B, both of which are reasonably likely to be sampled from their respective distributions. These are also both realistic possible values for the true conversion rate of each variant, but in this case, they indicate that variant B is actually worse than variant A.

We can imagine that the posterior distribution represents all the worlds that could exist based on our current state of beliefs regarding each conversion rate. Every time we sample from each distribution, we're seeing what one possible world could look like. We can tell visually in Figure 15-1 that we should expect more worlds where B is truly the better variant. The more frequently we sample, the more precisely we can tell in exactly how many worlds, of all the worlds we've sampled from, B is the better variant. Once we have our samples, we can look at the ratio of worlds where B is the best to the total number of worlds we've looked at and get an exact probability that B is in fact greater than A.

In How Many Worlds Is B the Better Variant?

Now we just have to write the code that will perform this sampling. R's rbeta() function allows us to automatically sample from a beta distribution. We can consider each comparison of two samples a single trial. The more trials we run, the more precise our result will be, so we'll start with 100,000 trials by assigning this value to the variable n.trials:

```
n.trials <- 100000
```

Next we'll put our prior alpha and beta values into variables:

```
prior.alpha <- 3
```

```
prior.beta <- 7
```

Then we need to collect samples from each variant. We'll use rbeta() for this:

```
a.samples <- rbeta(n.trials,36+prior.alpha,114+prior.beta)
b.samples <- rbeta(n.trials,50+prior.alpha,100+prior.beta)
```

We're saving the results of the rbeta() samples into variables, too, so we can access them more easily. For each variant, we input the number of people who clicked through to the blog and the number of people who didn't.

Finally, we compare how many times the b.samples are greater than the a.samples and divide that number by n.trials, which will give us the percentage of the total trials where variant B was greater than variant A:

```
p.b_superior <- sum(b.samples > a.samples)/n.trials
```

The result we end up with is:

```
p.b_superior
> 0.96
```

What we see here is that in 96 percent of the 100,000 trials, variant B was superior. We can imagine this as looking at 100,000 possible worlds. Based on the distribution of possible conversion rates for each variant, in 96 percent of the worlds variant B was the better of the two. This result shows that, even with a relatively small number of observed samples, we have a pretty strong belief that B is the better variant. If you've ever done *t*-tests in classical statistics, this is roughly equivalent—if we used a Beta(1,1) prior—to getting a *p*-value of 0.04 from a single-tailed *t*-test (often considered "statistically significant"). However, the beauty of our approach is that we were able to build this test from scratch using just our knowledge of probability and a straightforward simulation.

How Much Better Is Each Variant B Than Each Variant A?

Now we can say precisely how certain we are that B is the superior variant. However, if this email campaign were for a real business, simply saying "B is better" wouldn't be a very satisfactory answer. Don't you really want to know *how much better*?

This is the real power of our Monte Carlo simulation. We can take the exact results from our last simulation and test how much better variant B is likely to be by looking at how many times greater the B samples are than the A samples. In other words, we can look at this ratio:

$$\frac{\text{B samples}}{\text{A samples}}$$

In R, if we take the a.samples and b.samples from before, we can compute b.samples/a.samples. This will give us a distribution of the relative improvements from variant A to variant B. When we plot out this distribution as a histogram, as shown in Figure 15-3, we can see how much we expect variant B to improve our click-through rate.

From this histogram we can see that variant B will most likely be about a 40 percent improvement (ratio of 1.4) over A, although there is an entire range of possible values. As we discussed in Chapter 13, the cumulative distribution function (CDF) is much more useful than a histogram for reasoning about our results. Since we're working with data rather than a mathematical function, we'll compute the *empirical* cumulative distribution function with R's ecdf() function. The eCDF is illustrated in Figure 15-4.

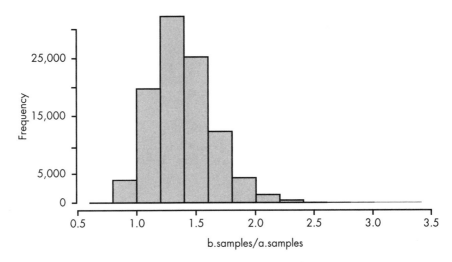

Figure 15-3: A histogram of possible improvements we might see

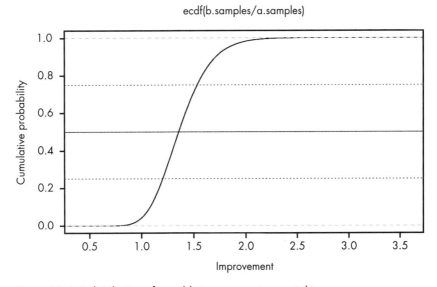

Figure 15-4: A distribution of possible improvements we might see

Now we can see our results more clearly. There is really just a small, small chance that A is better, and even if it is better, it's not going to be by much. We can also see that there's about a 25 percent chance that variant B is a 50 percent or more improvement over A, and even a reasonable chance it could be more than double the conversion rate! Now, in choosing B over A, we can actually reason about our risk by saying, "The chance that B is 20 percent worse is roughly the same that it's 100 percent better." Sounds like a good bet to me, and a much better statement of our knowledge than, "There is a statistically significant difference between B and A."

Wrapping Up

In this chapter we saw how parameter estimation naturally extends to a form of hypothesis testing. If the hypothesis we want to test is "variant B has a better conversion rate than variant A," we can start by first doing parameter estimation for the possible conversion rates of each variant. Once we know those estimates, we can use the Monte Carlo simulation in order to sample from them. By comparing these samples, we can come up with a probability that our hypothesis is true. Finally, we can take our test one step further by seeing how well our new variant performs in these possible worlds, estimating not only whether the hypothesis is true, but also how much improvement we are likely to see.

Exercises

Try answering the following questions to see how well you understand running A/B tests. The solutions can be found at *https://nostarch.com/learnbayes/*.

1. Suppose a director of marketing with many years of experience tells you he believes very strongly that the variant without images (B) won't perform any differently than the original variant. How could you account for this in our model? Implement this change and see how your final conclusions change as well.

2. The lead designer sees your results and insists that there's no way that variant B should perform better with no images. She feels that you should assume the conversion rate for variant B is closer to 20 percent than 30 percent. Implement a solution for this and again review the results of our analysis.

3. Assume that being 95 percent certain means that you're more or less "convinced" of a hypothesis. Also assume that there's no longer any limit to the number of emails you can send in your test. If the true conversion for A is 0.25 and for B is 0.3, explore how many samples it would take to convince the director of marketing that B was in fact superior. Explore the same for the lead designer. You can generate samples of conversions with the following snippet of R:

```
true.rate <- 0.25
number.of.samples <- 100
results <- runif(number.of.samples) <= true.rate
```

16

INTRODUCTION TO THE BAYES FACTOR AND POSTERIOR ODDS: THE COMPETITION OF IDEAS

In the previous chapter, we saw that we can view a hypothesis test as an extension of parameter estimation. In this chapter, we'll think about hypothesis tests instead as a way to compare ideas with an important mathematical tool called the *Bayes factor*. The Bayes factor is a formula that tests the plausibility of one hypothesis by comparing it to another. The result tells us how many times more likely one hypothesis is than the other.

We'll then see how to combine the Bayes factor with our prior beliefs to come up with the posterior odds, which tells us how much stronger one belief is than the other at explaining our data.

Revisiting Bayes' Theorem

Chapter 6 introduced Bayes' theorem, which takes the following form:

$$P(H \mid D) = \frac{P(H) \times P(D \mid H)}{P(D)}$$

Recall that there are three parts of this formula that have special names:

- $P(H \mid D)$ is the *posterior probability*, which tells us how strongly we should believe in our hypothesis, given our data.
- $P(H)$ is the *prior belief*, or the probability of our hypothesis prior to looking at the data.
- $P(D \mid H)$ is the *likelihood* of getting the existing data if our hypothesis were true.

The last piece, $P(D)$, is the probability of the data observed independent of the hypothesis. We need $P(D)$ in order to make sure that our posterior probability is correctly placed somewhere between 0 and 1. If we have all of these pieces of information, we can calculate exactly how strongly we should believe in our hypothesis given the data we've observed. But as I mentioned in Chapter 8, $P(D)$ is often very hard to define. In many cases, it's not obvious how we can figure out the probability of our data. $P(D)$ is also totally unnecessary if all we care about is comparing the relative strength of two different hypotheses.

For these reasons, we often use the *proportional form* of Bayes' theorem, which allows us to analyze the strength of our hypotheses without knowing $P(D)$. It looks like this:

$$P(H \mid D) \propto P(H) \times P(D \mid H)$$

In plain English, the proportional form of Bayes' theorem says that the posterior probability of our hypothesis is proportional to the prior multiplied by the likelihood. We can use this to compare two hypotheses by examining the ratio of the prior belief multiplied by the likelihood for each hypothesis using the *ratio of posteriors* formula:

$$\frac{P(H_1) \times P(D \mid H_1)}{P(H_2) \times P(D \mid H_2)}$$

What we have now is a ratio of how well each of our hypotheses explains the data we've observed. That is, if the ratio is 2, then H_1 explains the observed data twice as well as H_2, and if the ratio is 1/2, then H_2 explains the data twice as well as H_1.

Building a Hypothesis Test Using the Ratio of Posteriors

The ratio of posteriors formula gives us the *posterior odds*, which allows us to test hypotheses or beliefs we have about data. Even when we do know $P(D)$, the posterior odds is a useful tool because it allows us to compare ideas. To better understand the posterior odds, we'll break down the ratio of posteriors formula into two parts: the likelihood ratio, or the Bayes factor, and the ratio of prior probabilities. This is a standard, and very helpful, practice that makes it much easier to reason about the likelihood and the prior probability separately.

The Bayes Factor

Using the ratio of posteriors formula, let's assume that $P(H_1) = P(H_2)$—that is, that our prior belief in each hypothesis is the same. In that case, the ratio of prior beliefs in the hypotheses is just 1, so all that's left is:

$$\frac{P(D \mid H_1)}{P(D \mid H_2)}$$

This is the Bayes factor, the ratio between the likelihoods of two hypotheses.

Take a moment to really think about what this equation is saying. When we consider how we're going to argue for our H_1—that is, our belief about the world—we think about gathering evidence that supports our beliefs. A typical argument, therefore, involves building up a set of data, D_1, that supports H_1, and then arguing with a friend who has gathered a set of data, D_2, that supports their hypothesis, H_2.

In Bayesian reasoning, though, we're not gathering evidence to support our ideas; we're looking to see how well our ideas explain the evidence in front of us. What this ratio tells us is the likelihood of what we've seen given what *we* believe to be true compared to what *someone else* believes to be true. Our hypothesis wins when it explains the world better than the competing hypothesis.

If, however, the competing hypothesis explains the data much better than ours, it might be time to change our beliefs. The key here is that in Bayesian reasoning, we don't worry about supporting our beliefs—we are focused on how well our beliefs support the data we observe. In the end, data can either confirm our ideas or lead us to change our minds.

Prior Odds

So far we have assumed that the prior probability of each hypothesis is the same. This is clearly not always the case: a hypothesis may explain the data well even if it is very unlikely. If you've lost your phone, for example, both the belief that you left it in the bathroom and the belief that aliens took it to examine human technology explain the data quite well. However, the bathroom hypothesis is clearly much more likely. This is why we need to consider the ratio of prior probabilities:

$$\frac{P(H_1)}{P(H_2)}$$

This ratio compares the probability of two hypotheses before we look at the data. When used in relation to the Bayes factor, this ratio is called the *prior odds* in our H_1 and written as $O(H_1)$. This representation is helpful because it lets us easily note how strongly (or weakly) we believe in the hypothesis we're testing. When this number is greater than 1, it means the prior odds favor our hypothesis, and when it is a fraction less than 1, it means they're against our hypothesis. For example, $O(H_1) = 100$ means that, without any other information, we believe H_1 is 100 times more likely than the alternative hypothesis. On the other hand, when $O(H_1) = 1/100$, the alternative hypothesis is 100 times more likely than ours.

Posterior Odds

If we put together the Bayes factor and the prior odds, we get the posterior odds:

$$\text{posterior odds} = O(H_1)\frac{P(D \mid H_1)}{P(D \mid H_2)}$$

The posterior odds calculates how many times better our hypothesis explains the data than a competing hypothesis.

Table 16-1 lists some guidelines for evaluating various posterior odds values.

Table 16-1: Guidelines for Evaluating Posterior Odds

Posterior odds	Strength of evidence
1 to 3	Interesting, but nothing conclusive
3 to 20	Looks like we're on to something
20 to 150	Strong evidence in favor of H_1
> 150	Overwhelming evidence

We can look at the reciprocal of these odds to decide when to change our mind about an idea.

While these values can serve as a useful guide, Bayesian reasoning is still a form of reasoning, which means you have to use some judgment. If you're having a casual disagreement with a friend, a posterior odds of 2 might be enough to make you feel confident. If you're trying to figure out if you're drinking poison, a posterior odds of 100 still might not cut it.

Next, we'll look at two examples in which we use the Bayes factor to determine the strength of our beliefs.

Testing for a Loaded Die

We can use the Bayes factor and posterior odds as a form of hypothesis testing in which each test is a competition between two ideas. Suppose your friend has a bag with three six-sided dice in it, and one die is weighted so that it lands on 6 half the time. The other two are traditional dice whose probability of rolling a 6 is ⅙. Your friend pulls out a die and rolls 10 times, with the following results:

$$6, 1, 3, 6, 4, 5, 6, 1, 2, 6$$

We want to figure out if this is the loaded die or a regular die. We can call the loaded die H_1 and the regular die H_2.

Let's start by working out the Bayes factor:

$$\frac{P(D \mid H_1)}{P(D \mid H_2)}$$

The first step is calculating $P(D \mid H)$, or the likelihood of H_1 and H_2 given the data we've observed. In this example, your friend rolled four 6s and six non-6s. We know that if the die is loaded, the probability of rolling a 6 is 1/2 and the probability of rolling any non-6 is also 1/2. This means the likelihood of seeing this data given that we've used the loaded die is:

$$P(D \mid H_1) = \left(\frac{1}{2}\right)^4 \times \left(\frac{1}{2}\right)^6 = 0.00098$$

In the case of the fair die, the probability of rolling a 6 is 1/6, while the probability of rolling anything else is 5/6. This means our likelihood of seeing this data for H_2, the hypothesis that the die is fair, is:

$$P(D \mid H_2) = \left(\frac{1}{6}\right)^4 \times \left(\frac{5}{6}\right)^6 = 0.00026$$

Now we can compute our Bayes factor, which will tell us how much better H_1 is than H_2 at explaining our data, assuming each hypothesis was equally probable in the first place (meaning that the prior odds ratio is 1):

$$\frac{P(D \mid H_1)}{P(D \mid H_2)} = \frac{0.00098}{0.00026} = 3.77$$

This means that H_1, the belief that the die is loaded, explains the data we observed almost four times better than H_2.

However, this is true only if H_1 and H_2 are both just as likely to be true in the first place. But we know there are two fair dice in the bag and only one loaded die, which means that each hypothesis was *not* equally likely. Based on the distribution of the dice in the bag, we know that these are the prior probabilities for each hypothesis:

$$P(H_1) = \frac{1}{3}; P(H_2) = \frac{2}{3}$$

From these, we can calculate the prior odds for H_1:

$$\text{prior odds} = O(H_1) = \frac{P(H_1)}{P(H_2)} = \frac{\frac{1}{3}}{\frac{2}{3}} = \frac{1}{2}$$

Because there is only one loaded die in the bag and two fair dice, we're twice as likely to pull a fair die than a loaded one. With our prior odds for H_1, we can now compute our full posterior odds:

$$\text{posterior odds} = O(H_1) \times \frac{P(D \mid H_1)}{P(D \mid H_2)} = \frac{1}{2} \times 3.77 = 1.89$$

While the initial likelihood ratio showed that H_1 explained the data almost four times as well as H_2, the posterior odds shows us that, because H_1 is only half as likely as H_2, H_1 is actually only about twice as strong of an explanation as H_2.

From this, if you absolutely had to draw a conclusion about whether the die was loaded or not, your best bet would be to say that it is indeed loaded. However, a posterior odds of less than 2 is not particularly strong evidence in favor of H_1. If you really wanted to know whether or not the die was loaded, you would need to roll it a few more times until the evidence in favor of one hypothesis or the other was great enough for you to make a stronger decision.

Now let's look at a second example of using the Bayes factor to determine the strength of our beliefs.

Self-Diagnosing Rare Diseases Online

Many people have made the mistake of looking up their symptoms and ailments online late at night, only to find themselves glued to the screen in terror, sure they are the victim of some strange and terrible disease! Unfortunately for them, their analysis almost always excludes Bayesian reasoning, which might help alleviate some unnecessary anxiety. In this example, let's assume you've made the mistake of looking up your symptoms and have found two possible ailments that fit. Rather than panicking for no reason, you'll use posterior odds to weigh the odds of each.

Suppose you wake up one day with difficulty hearing and a ringing (tinnitus) in one ear. It annoys you all day, and when you get home from work, you decide it's high time to search the web for potential causes of your symptoms. You become increasingly concerned, and finally come to two possible hypotheses:

Earwax impaction   You have too much earwax in one ear. A quick visit to the doctor will clear up this condition.

Vestibular schwannoma   You have a brain tumor growing on the myelin sheath of the vestibular nerve, causing irreversible hearing loss and possibly requiring brain surgery.

Of the two, the possibility of vestibular schwannoma is the most worrying. Sure, it could be just earwax, but what if it's not? What if you *do* have a brain tumor? Since you're most worried about the possibility of a brain tumor, you decide to make this your H_1. Your H_2 is the hypothesis that you have too much earwax in one ear.

Let's see if posterior odds can calm you down.

As in our last example, we'll start our exploration by looking at the likelihood of observing these symptoms if each hypothesis were true, and compute the Bayes factor. This means we need to compute $P(D \mid H)$. You've observed two symptoms: hearing loss and tinnitus.

For vestibular schwannoma, the probability of experiencing hearing loss is 94 percent, and the probability of experiencing tinnitus is 83 percent, which means the probability of having hearing loss and tinnitus if you have vestibular schwannoma is:

$$P(D \mid H_1) = 0.94 \times 0.89 = 0.78$$

Next, we'll do the same for H_2. For earwax impaction, the probability of experiencing hearing loss is 63 percent, and the probability of experiencing tinnitus is 55 percent. The likelihood of having your symptoms if you have impacted earwax is:

$$P(D \mid H_2) = 0.63 \times 0.55 = 0.35$$

Now we have enough information to look at our Bayes factor:

$$\frac{P(D \mid H_1)}{P(D \mid H_2)} = \frac{0.78}{0.35} = 2.23$$

Yikes! Looking at just the Bayes factor doesn't do much to help alleviate your concerns of having a brain tumor. Taking only the likelihood ratio into account, it appears that you're more than twice as likely to experience these symptoms if you have vestibular schwannoma than if you have earwax impaction! Luckily, we're not done with our analysis yet.

The next step is to determine the prior odds of each hypothesis. Symptoms aside, how likely is it for someone to have one issue versus the other? We can find epidemiological data for each of these diseases. It turns out that vestibular schwannoma is a rare condition. Only 11 in 1,000,000 people contract it each year. The prior odds look like this:

$$P(H_1) = \frac{11}{1,000,000}$$

Unsurprisingly, earwax impaction is much, much more common, with 37,000 cases per 1,000,000 people in a year:

$$P(H_2) = \frac{37,000}{1,000,000}$$

To get the prior odds for H_1, we need to look at the ratio of these two prior probabilities:

$$O(H_1) = \frac{P(H_1)}{P(H_2)} = \frac{\dfrac{11}{1,000,000}}{\dfrac{37,000}{1,000,000}} = \frac{11}{37,000}$$

Based on prior information alone, a given person is about 3,700 times more likely to have an earwax impaction than vestibular schwannoma. But before you can breathe easy, we need to compute the full posterior odds. This just means multiplying our Bayes factor by our prior odds:

$$O(H_1) \times \frac{P(D \mid H_1)}{P(D \mid H_2)} = \frac{11}{37,000} \times 2.23 = \frac{223}{370,000}$$

This result shows that H_2 is about 1,659 times more likely than H_1. Finally, you can relax, knowing that a visit to the doctor in the morning for a simple ear cleaning will likely clear all this up!

In everyday reasoning, it's easy to overestimate the probability of scary situations, but by using Bayesian reasoning, we can break down the real risks and see how likely they actually are.

Wrapping Up

In this chapter, you learned how to use the Bayes factor and posterior odds to compare two hypotheses. Rather than focusing on providing data to support our beliefs, the Bayes factor tests how well our beliefs support the data we've observed. The result is a ratio that reflects how many times better one hypothesis explains the data than the other. We can use it to strengthen our prior beliefs when they explain the data better than alternative beliefs. On the other hand, when the result is a fraction, we might want to consider changing our minds.

Exercises

Try answering the following questions to see how well you understand the Bayes factor and posterior odds. The solutions can be found at *https://nostarch.com/learnbayes/*.

1. Returning to the dice problem, assume that your friend made a mistake and suddenly realized that there were, in fact, two loaded dice and only one fair die. How does this change the prior, and therefore the posterior odds, for our problem? Are you more willing to believe that the die being rolled is the loaded die?

2. Returning to the rare diseases example, suppose you go to the doctor, and after having your ears cleaned you notice that your symptoms persist. Even worse, you have a new symptom: vertigo. The doctor proposes another possible explanation, labyrinthitis, which is a viral infection of the inner ear in which 98 percent of cases involve vertigo. However, hearing loss and tinnitus are less common in this disease; hearing loss occurs only 30 percent of the time, and tinnitus occurs only 28 percent of the time. Vertigo is also a possible symptom of vestibular schwannoma, but occurs in only 49 percent of cases. In the general population, 35 people per million contract labyrinthitis annually. What is the posterior odds when you compare the hypothesis that you have labyrinthitis against the hypothesis that you have vestibular schwannoma?

17

BAYESIAN REASONING IN THE TWILIGHT ZONE

In Chapter 16, we used the Bayes factor and posterior odds to find out how many times better one hypothesis was than a competing one. But these tools of Bayesian reasoning can do even more than just compare ideas. In this chapter, we'll use the Bayes factor and posterior odds to quantify how much evidence it should take to convince someone of a hypothesis. We'll also see how to estimate the strength of someone else's prior belief in a certain hypothesis. We'll do all of this using a famous episode of the classic TV series *The Twilight Zone*.

Bayesian Reasoning in the Twilight Zone

One of my favorite episodes of *The Twilight Zone* is called "The Nick of Time." In this episode, a young, newly married couple, Don and Pat, wait in a small-town diner while a mechanic repairs their car. In the diner, they come across a fortune-telling machine called the Mystic Seer that accepts yes or no questions and, for a penny, spits out cards with answers to each question.

Don, who is very superstitious, asks the Mystic Seer a series of questions. When the machine answers correctly, he begins to believe in its supernatural powers. However, Pat remains skeptical of the machine's powers, even as the Seer continues to provide correct answers.

Although Don and Pat are looking at the same data, they come to different conclusions. How can we explain why they reason differently when given the same evidence? We can use the Bayes factor to get deeper insight into how these two characters are thinking about the data.

Using the Bayes Factor to Understand the Mystic Seer

In the episode, we are faced with two competing hypotheses. Let's call them H and $\bar{H}$ (or "not H"), since one hypothesis is the negation of the other:

H The Mystic Seer truly can predict the future.

$\bar{H}$ The Mystic Seer just got lucky.

Our data, D, in this case is the sequence of n correct answers the Mystic Seer provides. The greater n is, the stronger the evidence in favor of H. The major assumption in the *Twilight Zone* episode is that the Mystic Seer *is* correct every time, so the question is: is this result supernatural, or is it merely a coincidence? For us, D, our data, always represents a sequence of n correct answers. Now we can assess our likelihoods, or the probability of getting our data given each hypothesis.

$P(D \mid H)$ is the probability of getting n correct answers in a row given that the Mystic Seer can predict the future. This likelihood will always be 1, no matter the number of questions asked. This is because, if the Mystic Seer is supernatural, it will always pick the right answer, whether it is asked one question or a thousand. Of course, this also means that if the Mystic Seer gets a single answer wrong, the probability for this hypothesis will drop to 0, because a psychic machine wouldn't ever guess incorrectly. In that case, we might want to come up with a weaker hypothesis—for example, that the Mystic Seer is correct 90 percent of the time (we'll explore a similar problem in Chapter 19).

$P(D \mid \bar{H})$ is the probability of getting n correct answers in a row if the Mystic Seer is randomly spitting out answers. Here, $P(D \mid \bar{H})$ is 0.5^n. In other words, if the machine is just guessing, then each answer has a 0.5 chance of being correct.

To compare these hypotheses, let's look at the ratio of the two likelihoods:

$$\frac{P(D \mid H)}{P\left(D \mid \overline{H}\right)}$$

As a reminder, this ratio measures how many times more likely the data is, given H as opposed to $\overline{H}$, when we assume both hypotheses are equally likely. Now let's see how these ideas compare.

Measuring the Bayes Factor

As we did in the preceding chapter, we'll temporarily ignore the ratio of our prior odds and concentrate on comparing the ratio of the likelihoods, or the Bayes factor. We're assuming (for the time being) that the Mystic Seer has an equal chance of being supernatural as it does of being simply lucky.

In this example, our numerator, $P(D \mid H)$, is always 1, so for any value of n we have:

$$BF = \frac{P\left(D_n \mid H\right)}{P\left(D_n \mid \overline{H}\right)} = \frac{1}{0.5^n}$$

Let's imagine the Mystic Seer has given three correct answers so far. At this point, $P(D_3 \mid H) = 1$, and $P(D \mid H) = 0.5^3 = 0.125$. Clearly H explains the data better, but certainly nobody—not even superstitious Don—will be convinced by only three correct guesses. Assuming the prior odds are the same, our Bayes factor for three questions is:

$$BF = \frac{1}{0.125} = 8$$

We can use the same guidelines we used for evaluating posterior odds in Table 16-1 to evaluate Bayes factors here (if we assume each hypothesis is equally likely), as shown in Table 17-1. As you can see, a Bayes factor (BF) of 8 is far from conclusive.

Table 17-1: Guidelines for Evaluating Bayes Factors

BF	Strength of evidence
1 to 3	Interesting, but nothing conclusive
3 to 20	Looks like we're on to something
20 to 150	Strong evidence in favor of H_1
> 150	Overwhelming evidence in favor of H_1

So, at three questions answered correctly and with BF = 8, we should at least be curious about the power of the Mystic Seer, though we shouldn't be convinced yet.

But by this point in the episode, Don already seems pretty sure that the Mystic Seer is psychic. It takes only four correct answers for him to feel certain of it. On the other hand, it takes 14 questions for Pat to even *start considering* the possibility seriously, resulting in a Bayes factor of 16,384—way more evidence than she should need.

Calculating the Bayes factor doesn't explain why Don and Pat form different beliefs about the evidence, though. What's going on there?

Accounting for Prior Beliefs

The element missing in our model is each character's prior belief in the hypotheses. Remember that Don is extremely superstitious, while Pat is a skeptic. Clearly, Don and Pat are using extra information in their mental models, because each of them arrives at a conclusion of a different strength, and at very different times. This is fairly common in everyday reasoning: two people often respond differently to the exact same facts.

We can model this phenomenon by simply imagining the initial odds of $P(H)$ and $P(\overline{H})$ given no additional information. We call this the *prior odds ratio*, as you saw in Chapter 16:

$$\text{prior odds} = O(H) = \frac{P(H)}{P(\overline{H})}$$

The concept of prior beliefs in relation to the Bayes factor is actually pretty intuitive. Say we walk into the diner from *The Twilight Zone*, and I ask you, "What are the odds that the Mystic Seer is psychic?" You might reply, "Uh, one in a million! There's no way that thing is supernatural." Mathematically, we can express this as:

$$O(H) = \frac{1}{1,000,000}$$

Now let's combine this prior belief with our data. To do this, we'll multiply our prior odds with the results of the likelihood ratio to get our posterior odds for the hypothesis, given the data we've observed:

$$\text{posterior odds} = O(H \mid D) = O(H) \times \frac{P(D \mid H)}{P(D \mid \overline{H})}$$

Thinking there's only a one in a million chance the Mystic Seer is psychic before looking at any evidence is pretty strong skepticism. The Bayesian approach reflects this skepticism quite well. If you think the hypothesis that the Mystic Seer is supernatural is extremely unlikely from the start, then you'll require significantly more data to be convinced otherwise. Suppose the Mystic Seer gets five answers correct. Our Bayes factor then becomes:

$$\text{BF} = \frac{1}{0.5^5} = 32$$

A Bayes factor of 32 is a reasonably strong belief that the Mystic Seer is truly supernatural. However, if we add in our very skeptical prior odds to calculate our posterior odds, we get the following results:

$$\text{posterior odds} = O(H \mid D_5) \times \frac{P(D_5 \mid H)}{P(D_5 \mid \overline{H})} = \frac{1}{1,000,000} \times \frac{1}{0.5^5} = 0.000032$$

Now our posterior odds tell us it's extremely unlikely that the machine is psychic. This result corresponds quite well with our intuition. Again, if you really don't believe in a hypothesis from the start, it's going to take a lot of evidence to convince you otherwise.

In fact, if we work backward, posterior odds can help us figure out how much evidence we'd need to make you believe H. At a posterior odds of 2, you'd just be starting to consider the supernatural hypothesis. So, if we solve for a posterior odds of greater than 2, we can determine what it would take to convince you.

$$\frac{1}{1,000,000} \times \frac{1}{0.5^n} > 2$$

If we solve for n to the nearest whole number, we get:

$$n > 21$$

At 21 correct answers in a row, even a strong skeptic should start to think that the Seer may, in fact, be psychic.

Thus, our prior odds can do much more than tell us how strongly we believe something given our background. It can also help us quantify exactly how much evidence we would need to be convinced of a hypothesis. The reverse is true, too; if, after 21 correct answers in a row, you find yourself believing strongly in H, you might want to weaken your prior odds.

Developing Our Own Psychic Powers

At this point, we've learned how to compare hypotheses and calculate how much favorable evidence it would take to convince us of H, given our prior belief in H. Now we'll look at one more trick we can do with posterior odds: quantifying Don and Pat's prior beliefs based on their reactions to the evidence.

We don't know exactly how strongly Don and Pat believe in the possibility that the Mystic Seer is psychic when they first walk into the diner. But we *do* know it takes Don about seven correct questions to become essentially certain of the Mystic Seer's supernatural abilities. We can estimate that at this point Don's posterior odds are 150—the threshold for *very strong* beliefs, according to Table 17-1. Now we can write out everything we know, except for $O(H)$, which we'll be solving for:

$$150 = O(H) \times \frac{P(D_7 \mid H)}{P(D_7 \mid \overline{H})} = O(H) \times \frac{1}{0.5^7}$$

Solving this for $O(H)$ gives us:

$$O(H)_{\text{Don}} = 1.17$$

What we have now is a quantitative model for Don's superstitious beliefs. Because his initial odds ratio is greater than 1, Don walks into the diner being slightly more willing than not to believe that the Mystic Seer is supernatural, before collecting any data at all. This makes sense, of course, given his superstitious nature.

Now on to Pat. At around 14 correct answers, Pat grows nervous, calling the Mystic Seer "a stupid piece of junk!" Although she has begun to suspect that the Mystic Seer might be psychic, she's not nearly as certain as Don. I would estimate that her posterior odds are 5—the point at which she might start thinking, "Maybe the Mystic Seer *could have* psychic powers . . ." Now we can create the posterior odds for Pat's beliefs in the same way:

$$5 = O(H) \times \frac{P(D_{14} \mid H)}{P(D_{14} \mid \overline{H})} = O(H) \times \frac{1}{0.5^{14}}$$

When we solve for $O(H)$, we can model Pat's skepticism as:

$$O(H)_{\text{Pat}} = 0.0003$$

In other words, Pat, walking into the diner, would claim that the Seer has about a 1 in 3,000 chance of being supernatural. Again, this corresponds to our intuition; Pat begins with the very strong belief that the fortune-telling machine is nothing more than a fun game to play while she and Don wait for food.

What we've done here is remarkable. We've used our rules of probability to come up with a quantitative statement about what someone believes. In essence, we have become mind readers!

Wrapping Up

In this chapter, we explored three ways of using Bayes factors and posterior odds in order to reason about problems probabilistically. We started by revisiting what we learned in the previous chapter: that we can use posterior odds as a way to compare two ideas. Then we saw that if we know our prior belief in the odds of one hypothesis versus another, we can calculate exactly how much evidence it will take to convince us that we should change our beliefs. Finally, we used posterior odds to assign a value for each person's prior beliefs by looking at how much evidence it takes to convince them. In the end, posterior odds is far more than just a way to test ideas. It provides us with a framework for thinking about reasoning under uncertainty.

You can now use your own "mystic" powers of Bayesian reasoning to answer the exercises below:

Exercises

Try answering the following questions to see how well you understand quantifying the amount of evidence it should take to convince someone of a hypothesis and estimating the strength of someone else's prior belief. The solutions can be found at *https://nostarch.com/learnbayes/*.

1. Every time you and your friend get together to watch movies, you flip a coin to determine who gets to choose the movie. Your friend always picks heads, and every Friday for 10 weeks, the coin lands on heads. You develop a hypothesis that the coin has two heads sides, rather than both a heads side and a tails side. Set up a Bayes factor for the hypothesis that the coin is a trick coin over the hypothesis that the coin is fair. What does this ratio alone suggest about whether or not your friend is cheating you?

2. Now imagine three cases: that your friend is a bit of a prankster, that your friend is honest most of the time but can occasionally be sneaky, and that your friend is very trustworthy. In each case, estimate some prior odds ratios for your hypothesis and compute the posterior odds.

3. Suppose you trust this friend deeply. Make the prior odds of them cheating 1/10,000. How many times would the coin have to land on heads before you feel unsure about their innocence—say, a posterior odds of 1?

4. Another friend of yours also hangs out with this same friend and, after only four weeks of the coin landing on heads, feels certain you're both being cheated. This confidence implies a posterior odds of about 100. What value would you assign to this other friend's prior belief that the first friend is a cheater?

18

WHEN DATA DOESN'T CONVINCE YOU

 In the previous chapter, we used Bayesian reasoning to reason about two hypotheses from an episode of *The Twilight Zone*:

- H The fortune-telling Mystic Seer is supernatural.
- $\bar{H}$ The fortune-telling Mystic Seer isn't supernatural, just lucky.

We also learned how to account for skepticism by changing the prior odds ratio. For example, if you, like me, believe that the Mystic Seer definitely isn't psychic, then you might want to set the prior odds extremely low—something like 1/1,000,000.

However, depending on your level of personal skepticism, you might feel that even a 1/1,000,000 odds ratio wouldn't be quite enough to convince you of the seer's power.

Maybe even after receiving 1,000 correct answers from the seer—which, despite your very skeptical prior odds, would suggest you were astronomically in favor of believing the seer is psychic—you still wouldn't buy into its supernatural powers. We could represent this by simply making our prior odds even more extreme, but I personally don't find this solution very satisfying because no amount of data would convince me that the Mystic Seer is, in fact, psychic.

In this chapter, we'll take a deeper look at problems where the data doesn't convince people in the way we expect it to. In the real world, these situations are fairly common. Anyone who has argued with a relative over a holiday dinner has likely noticed that oftentimes the more contradictory evidence you give, the more they seem to be convinced of their preexisting belief! In order to fully understand Bayesian reasoning, we need to be able to understand, mathematically, why situations like these arise. This will help us identify and avoid them in our statistical analysis.

A Psychic Friend Rolling Dice

Suppose your friend tells you they can predict the outcome of a six-sided die roll with 90 percent accuracy because they are psychic. You find this claim difficult to believe, so you set up a hypothesis test using the Bayes factor. As in the Mystic Seer example, you have two hypotheses you want to compare:

$$H_1 : P(\text{correct}) = \frac{1}{6} \quad H_2 : P(\text{correct}) = \frac{9}{10}$$

The first hypothesis, H_1, represents your belief that the die is fair, and that your friend is not psychic. If the die is fair, there is a 1 in 6 chance of guessing the result correctly. The second hypothesis, H_2, represents your friend's belief that they can, in fact, predict the outcome of a die roll 90 percent of the time and is therefore given a 9/10 ratio. Next we need some data to start testing their claim. Your friend rolls the die 10 times and correctly guesses the outcome of the roll 9 times.

Comparing Likelihoods

As we often have in previous chapters, we'll start by looking at the Bayes factor, assuming for now that the prior odds for each hypothesis are equal. We'll formulate our likelihood ratio as:

$$\frac{P(D \mid H_2)}{P(D \mid H_1)}$$

so that our results will tell us how many times better (or worse) your friend's claim of being psychic explains the data than your hypothesis does. For this example, we'll use the variable BF for "Bayes factor" in our equations for brevity. Here is our result, taking into account the fact that your friend correctly predicted 9 out of 10 rolls:

$$BF = \frac{P(D_{10} \mid H_2)}{P(D_{10} \mid H_1)} = \frac{\left(\frac{9}{10}\right)^9 \times \left(1 - \frac{9}{10}\right)^1}{\left(\frac{1}{6}\right)^9 \times \left(1 - \frac{1}{6}\right)^1} = 468,517$$

Our likelihood ratio shows that the friend-being-psychic hypothesis explains the data 468,517 times better than the hypothesis that your friend is just lucky. This is a bit concerning. According to the Bayes factor chart we saw in earlier chapters, this means we should be nearly certain that H_2 is true and your friend is psychic. Unless you're already a deep believer in the possibility of psychic powers, something seems very wrong here.

Incorporating Prior Odds

In most cases in this book where the likelihood alone gives us strange results, we can solve the problem by including our prior probabilities. Clearly, we don't believe in our friend's hypothesis nearly as strongly as we believe in our own, so it makes sense to create a strong prior odds in favor of our hypothesis. We can start by simply setting our odds ratio high enough that it cancels out the extreme result of the Bayes factor, and see if this fixes our problem:

$$O(H_2) = \frac{1}{468,517}$$

Now, when we work out our full posterior odds, we find that we are, once again, unconvinced that your friend is psychic:

$$\text{posterior} = O(H_2) \times \frac{P(D_{10} \mid H_2)}{P(D_{10} \mid H_1)} = 1$$

For now, it looks like prior odds have once again saved us from a problem that occurred when we looked only at the Bayes factor.

But suppose your friend rolls the die five more times and successfully predicts all five outcomes. Now we have a new set of data, D_{15}, which represents 15 rolls of a die, 14 of which your friend guessed accurately. Now when we calculate our posterior odds, we see that even our extreme prior is of little help:

$$\text{posterior} = O(H_2) \times \frac{P(D_{15} \mid H_2)}{P(D_{15} \mid H_1)} = \frac{1}{468,517} \times \frac{\left(\frac{9}{10}\right)^{14} \times \left(1 - \frac{9}{10}\right)^1}{\left(\frac{1}{6}\right)^{14} \times \left(1 - \frac{1}{6}\right)^1} = 4,592$$

Using our existing prior, with just five more rolls of the die, we have posterior odds of 4,592—which means we're back to being nearly certain that your friend is truly psychic!

In most of our previous problems, we've corrected nonintuitive posterior results by adding a sane prior. We've added a pretty extreme prior against your friend being psychic, but our posterior odds are still strongly in favor of the hypothesis that they're psychic.

This is a major problem, because Bayesian reasoning should align with our everyday sense of logic. Clearly, 15 rolls of a die with 14 successful guesses is highly unusual, but it's unlikely to convince many people that the guesser truly possesses psychic powers! However, if we can't explain what's going on here with our hypothesis test, it means that we really can't rely on our test to solve our everyday statistical problems.

Considering Alternative Hypotheses

The issue here is that we *don't want to believe your friend is psychic*. If you found yourself in this situation in real life, it's likely you would quickly come to some alternative conclusion. You might come to believe that your friend is using a loaded die that rolls a certain value about 90 percent of the time, for example. This represents a *third* hypothesis. Our Bayes factor is looking at only two possible hypotheses: H_1, the hypothesis that the die is fair, and H_2, the hypothesis that your friend is psychic.

Our Bayes factor so far tells us that it's far more likely that our friend is psychic than that they are guessing the rolls of a fair die correctly. When we think of the conclusion in those terms, it makes more sense: with these results, it's extremely unlikely that the die is fair. We don't feel comfortable accepting the H_2 alternative, because our own beliefs about the world don't support the idea that H_2 is a realistic explanation.

It's important to understand that a hypothesis test compares only two explanations for an event, but very often there are countless possible explanations. If the winning hypothesis doesn't convince you, you could always consider a third one.

Let's look at what happens when we compare H_2, our winning hypothesis, with a new hypothesis, H_3: that the die is rigged so it has a certain outcome 90 percent of the time.

We'll start with a new prior odds about H_2, which we'll call $O(H_2)'$ (the tick mark is a common notation in mathematics meaning "like but not the same as"). This will represent the odds of H_2/H_3. For now, we'll just say that we believe it's 1,000 times more likely that your friend is using a loaded die than that your friend is really psychic (though our real prior might be much more extreme). That means the prior odds of your friend being psychic is 1/1,000. If we reexamine our new posterior odds, we get the following interesting result:

$$BF = O(H_2)' \times \frac{P(D_{15} \mid H_2)}{P(D_{15} \mid H_3)} = \frac{1}{1,000} \times \frac{\left(\frac{9}{10}\right)^{14} \times \left(1 - \frac{9}{10}\right)^{1}}{\left(\frac{9}{10}\right)^{14} \times \left(1 - \frac{9}{10}\right)^{1}} = \frac{1}{1,000}$$

According to this calculation, our posterior odds are the same as our prior odds, $O(H_2)'$. This happens because our two likelihoods are the same. In other words, $P(D_{15} \mid H_2) = P(D_{15} \mid H_3)$. For both hypotheses, the likelihood of your friend correctly guessing the outcome of the die roll is exactly the same for the loaded die because the probability each assigns to success is the same. This means that our Bayes factor will always be 1.

These results correspond quite well to our everyday intuition; after all, prior odds aside, each hypothesis explains the data we've seen equally well. That means that if, before considering the data, we believe one explanation is far more likely than the other, then no amount of new evidence will change our minds. So we no longer have a problem with the data we observed; we've simply found a better explanation for it.

In this scenario, no amount of data will change our mind about believing H_3 over H_2 because both explain what we've observed equally well, and we already think that H_3 is a far more likely explanation than H_2. What's interesting here is that we can find ourselves in this situation even if our prior beliefs are entirely irrational. Maybe you're a strong believer in psychic phenomena and think that your friend is the most honest person on earth. In this case, you might make the prior odds $O(H_2)' = 1,000$. If you believed this, no amount of data could convince you that your friend is using a loaded die.

In cases like this, it's important to realize that if you want to solve a problem, you need to be willing to change your prior beliefs. If you're unwilling to let go of unjustifiable prior beliefs, then, at the very least, you must acknowledge that you're no longer reasoning in a Bayesian—or logical—way at all. We all hold irrational beliefs, and that's perfectly okay, so long as we don't attempt to use Bayesian reasoning to justify them.

Arguing with Relatives and Conspiracy Theorists

Anyone who has argued with relatives over a holiday dinner about politics, climate change, or their favorite movies has experienced firsthand a situation in which they are comparing two hypotheses that both explain the data equally well (to the person arguing), and only the prior remains. How can we change someone else's (or our own) beliefs even when more data doesn't change anything?

We've already seen that if you compare the belief that your friend has a loaded die and the belief that they are psychic, more data will do nothing to change your beliefs about your friend's claim. This is because both your hypothesis and your friend's hypothesis explain the data equally well. In order for your friend to convince you that they are psychic, they have to alter your prior beliefs. For example, since you're suspicious that the die might be loaded, your friend could then offer to let you choose the die they roll. If you bought a new die and gave it to your friend, and they continued to accurately predict their rolls, you might start to be convinced. This same logic holds anytime you run into a problem where two hypotheses equally explain the data. In these cases, you must then see if there's anything you can change in your prior.

Suppose after you purchase the new die for your friend and they continue to succeed, you *still* don't believe them; you now claim that they must have a secret way of rolling. In response, your friend lets you roll the die for them, and they continue to successfully predict the rolls—yet you *still* don't believe them. In this scenario, something else is happening beyond just a hidden hypothesis. You now have an H_4—that your friend is completely cheating—and you won't change your mind. This means that for any D_n, $P(D_n \mid H_4) = 1$. Clearly we're out of Bayesian territory since you've essentially conceded that you won't change your mind, but let's see what happens mathematically if your friend persists in trying to convince you.

Let's look at how these two explanations, H_2 and H_4, compete using our data D_{10} with 9 correct predictions and 1 missed prediction. The Bayes factor for this is:

$$BF = \frac{P(D_{10} \mid H_2)}{P(D_{10} \mid H_4)} = \frac{\left(\dfrac{9}{10}\right)^9 \times \left(1 - \dfrac{9}{10}\right)^1}{1} = \frac{1}{26}$$

Because you refuse to believe anything other than that your friend is cheating, the probability of what you observe is, and will always be, 1. Even though the data is exactly as we would expect in the case of your friend being psychic, we find our beliefs explain the data 26 times as well. Your friend, deeply determined to change your stubborn mind, persists and rolls 100 times, getting 90 guesses right and 10 wrong. Our Bayes factor shows something very strange that happens:

$$BF = \frac{P(D_{100} \mid H_2)}{P(D_{100} \mid H_4)} = \frac{\left(\dfrac{9}{10}\right)^{90} \times \left(1 - \dfrac{9}{10}\right)^{10}}{1} = \frac{1}{131,272,619,177,803}$$

Even though the data seems to strongly support your friend's hypothesis, because you refuse to budge in your beliefs, you're now even more wildly convinced that you're right! When we don't allow our minds to be changed at all, more data only further convinces us we are correct.

This pattern may seem familiar to anyone who has argued with a politically radical relative or someone who adamantly believes in a conspiracy theory. In Bayesian reasoning, it is vital that our beliefs are at least falsifiable. In traditional science, *falsifiability* means that something can be disproved, but in our case it just means there has to be some way to reduce our belief in a hypothesis.

The danger of nonfalsifiable beliefs in Bayesian reasoning isn't just that they can't be proved wrong—it's that they are strengthened even by evidence that seems to contradict them. Rather than persisting in trying to convince you, your friend should have first asked, "What can I show you that would change your mind?" If your reply had been that *nothing* could change your mind, then your friend would be better off not presenting you with more evidence.

So, the next time you argue with a relative over politics or conspiracy theories, you should ask them: "What evidence would change your mind?" If they have no answer to this, you're better off not trying to defend your views with more evidence, as it will only increase your relative's certainty in their belief.

Wrapping Up

In this chapter, you learned about a few ways hypothesis tests can go wrong. Although the Bayes factor is a competition between two ideas, it's quite possible that there are other, equally valid, hypotheses worth testing out.

Other times, we find that two hypotheses explain the data equally well; you're just as likely to see your friend's correct predictions if they were caused by your friend's psychic ability or a trick in the die. When this is the case, only the prior odds ratio for each hypothesis matters. This also means that acquiring more data in those situations will never change our beliefs, because it will never give either hypothesis an edge over the other. In these cases, it's best to consider how you can alter the prior beliefs that are affecting the results.

In more extreme cases, we might have a hypothesis that simply refuses to be changed. This is like having a conspiracy theory about the data. When this is the case, not only will more data never convince us to change our beliefs, but it will actually have the opposite effect. If a hypothesis is not falsifiable, more data will only serve to make us more certain of the conspiracy.

Exercises

Try answering the following questions to see how well you understand how to deal with extreme cases in Bayesian reasoning. The solutions can be found at *https://nostarch.com/learnbayes/*.

1. When two hypotheses explain the data equally well, one way to change our minds is to see if we can attack the prior probability. What are some factors that might increase your prior belief in your friend's psychic powers?

2. An experiment claims that when people hear the word *Florida*, they think of the elderly and this has an impact on their walking speed. To test this, we have two groups of 15 students walk across a room; one group hears the word *Florida* and one does not. Assume H_1 = the groups don't move at different speeds, and H_2 = the Florida group is slower because of hearing the word *Florida*. Also assume:

$$BF = \frac{P(D \mid H_2)}{P(D \mid H_1)}$$

The experiment shows that H_2 has a Bayes factor of 19. Suppose someone is unconvinced by this experiment because H_2 had a lower prior odds. What prior odds would explain someone being unconvinced and what would the BF need to be to bring the posterior odds to 50 for this unconvinced person?

Now suppose the prior odds do not change the skeptic's mind. Think of an alternate H_3 that explains the observation that the Florida group is slower. Remember if H_2 and H_3 both explain the data equally well, only prior odds in favor of H_3 would lead someone to claim H_3 is true over H_2, so we need to rethink the experiment so that these odds are decreased. Come up with an experiment that could change the prior odds in H_3 over H_2.

19

FROM HYPOTHESIS TESTING TO PARAMETER ESTIMATION

So far, we've used posterior odds to compare only two hypotheses. That's fine for simple problems; even if we have three or four hypotheses, we can test them all by conducting multiple hypothesis tests, as we did in the previous chapter. But sometimes we want to search a really large space of possible hypotheses to explain our data. For example, you might want to guess how many jelly beans are in a jar, the height of a faraway building, or the exact number of minutes it will take for a flight to arrive. In all these cases, there are many, many possible hypotheses—too many to conduct hypothesis tests for all of them.

Luckily, there's a technique for handling this scenario. In Chapter 15, we learned how to turn a parameter estimation problem into a hypothesis test. In this chapter, we're going to do the opposite: by looking at a virtually continuous range of possible hypotheses, we can use the Bayes factor and posterior odds (a hypothesis test) as a form of parameter estimation! This approach allows us to evaluate more than just two hypotheses and provides us with a simple framework for estimating any parameter.

Is the Carnival Game Really Fair?

Suppose you're at a carnival. While walking through the games, you notice someone arguing with a carnival attendant near a pool of little plastic ducks. Curious, you get closer and hear the player yelling, "This game is rigged! You said there was a 1 in 2 chance of getting a prize and I've picked up 20 ducks and only received one prize! It looks to me like the chance of getting a prize is only 1 in 20!"

Now that you have a strong understanding of probability, you decide to settle this argument yourself. You explain to the attendant and the angry customer that if you observe some more games that day, you'll be able to use the Bayes factor to determine who's right. You decide to break up the results into two hypotheses: H_1, which represents the attendant's claim that the probability of a prize is 1/2, and H_2, the angry customer's claim that the probability of a prize is just 1/20:

$$H_1 : P(\text{prize}) = \frac{1}{2}$$

$$H_2 : P(\text{prize}) = \frac{1}{20}$$

The attendant argues that because he didn't watch the customer pick up ducks, he doesn't think you should use his reported data, since no one else can verify it. This seems fair to you. You decide to watch the next 100 games and use that as your data instead. After the customer has picked up 100 ducks, you observe that 24 of them came with prizes.

Now, on to the Bayes factor! Since we don't have a strong opinion about the claim from either the customer or the attendant, we won't worry about the prior odds or calculating our full posterior odds yet.

To get our Bayes factor, we need to compute $P(D|H)$ for each hypothesis:

$$P(D \mid H_1) = (0.5)^{24} \times (1 - 0.5)^{76}$$

$$P(D \mid H_2) = (0.05)^{24} \times (1 - 0.05)^{76}$$

Now, individually, both of these probabilities are quite small, but all we care about is the ratio. We'll look at our ratio in terms of H_2/H_1 so that our result will tell us how many times better the customer's hypothesis explains the data than the attendant's:

$$\frac{P(D \mid H_2)}{P(D \mid H_1)} = \frac{1}{653}$$

Our Bayes factor tells us that H_1, the attendant's hypothesis, explains the data 653 times as well as H_2, which means that the attendant's hypothesis (that the probability of getting a prize when picking up a duck is 0.5) is the more likely one.

This should immediately seem strange. Clearly, the probability of getting only 24 prizes out of a total of 100 ducks seems really unlikely if the true probability of a prize is 0.5. We can use R's pbinom() function (introduced in Chapter 13) to calculate the binomial distribution, which will tell us the probability of seeing 24 *or fewer* prizes, assuming that the probability of getting a prize is really 0.5:

```
> pbinom(24,100,0.5)
9.050013e-08
```

As you can see, the probability of getting 24 or fewer prizes if the true probability of a prize is 0.5 is extremely low; expanding it out to the full decimal values, we get a probability of 0.00000009050013! Something is definitely up with H_1. Even though we don't believe the attendant's hypothesis, it still explains the data much better than the customer's.

So what's missing? In the past, we've often found that the prior probability usually matters a lot when the Bayes factor alone doesn't give us an answer that makes sense. But as we saw in Chapter 18, there are cases in which the prior isn't the root cause of our problem. In this case, using the following equation seems reasonable, since we don't have a strong opinion either way:

$$O\left(\frac{H_2}{H_1}\right) = 1$$

But maybe the problem here is that you have a preexisting mistrust in carnival games. Because the result of the Bayes factor favors the attendant's hypothesis so strongly, we'd need our prior odds to be at least 653 to get a posterior odds that favors the customer's hypothesis:

$$O\left(\frac{H_2}{H_1}\right) = 653$$

That's a really deep distrust of the fairness of the game! There must be some problem here other than the prior.

Considering Multiple Hypotheses

One obvious problem is that, while it seems intuitively clear that the attendant is wrong in his hypothesis, the customer's alternative hypothesis is just too extreme to be right, either, so we have two wrong hypotheses. What if the customer thought the probability of winning was 0.2, rather than 0.05? We'll call this hypothesis H_3. Testing H_3 against the attendant's hypothesis radically changes the results of our likelihood ratio:

$$bf = \frac{P(D \mid H_3)}{P(D \mid H_1)} = \frac{(0.2)^{24} \times (1-0.2)^{76}}{(0.5)^{24} \times (1-0.5)^{76}} = 917{,}399$$

Here we see that H_3 explains the data wildly better than H_1. With a Bayes factor of 917,399, we can be certain that H_1 is far from the best hypothesis for explaining the data we've observed, because H_3 blows it out of the water. The trouble we had in our first hypothesis test was that the customer's belief was a far worse description of the event than the attendant's belief. As we can see, though, that doesn't mean the attendant was right. When we came up with an alternative hypothesis, we saw that it was a much better guess than either the attendant's or the customer's.

Of course, we haven't really solved our problem. What if there's an even better hypothesis out there?

Searching for More Hypotheses with R

We want a more general solution that searches all of our possible hypotheses and picks out the best one. To do this, we can use R's seq() function to create a sequence of hypotheses we want to compare to our H_1.

We'll consider every increment of 0.01 between 0 and 1 as a possible hypothesis. That means we'll consider 0.01, 0.02, 0.03, and so on. We'll call 0.01—the amount we're increasing each hypothesis by—dx (a common notation from calculus representing the "smallest change") and use it to define a hypotheses variable, which represents all of the possible hypotheses we want to consider. Here we use R's seq() function to generate a range of values for each hypothesis between 0 and 1 by incrementing the values by our dx:

```
dx <- 0.01
hypotheses <- seq(0,1,by=dx)
```

Next, we need a function that can calculate our likelihood ratio for any two hypotheses. Our bayes.factor() function will take two arguments: h_top, which is the probability of getting a prize for the hypothesis on the top (the numerator) and h_bottom, which is the hypothesis we're competing against (the attendant's hypothesis). We set this up like so:

```
bayes.factor <- function(h_top,h_bottom){
  ((h_top)^24*(1-h_top)^76)/((h_bottom)^24*(1-h_bottom)^76)
}
```

Finally, we compute the likelihood ratio for all of these possible hypotheses:

```
bfs <- bayes.factor(hypotheses,0.5)
```

Then, we use R's base plotting functionality to see what these likelihood ratios look like:

```
plot(hypotheses,bfs, type='l')
```

Figure 19-1 shows the resulting plot.

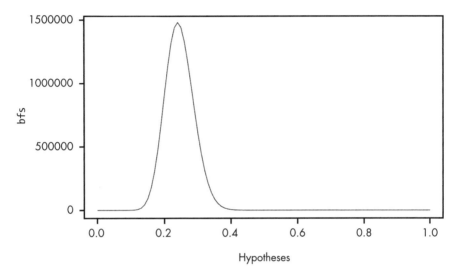

Figure 19-1: Plotting the Bayes factor for each of our hypotheses

Now we can see a clear distribution of different explanations for the data we've observed. Using R, we can look at a wide range of possible hypotheses, where each point in our line represents the Bayes factor for the corresponding hypothesis on the x-axis.

We can also see how high the largest Bayes factor is by using the max() function with our vector of bfs:

```
> max(bfs)
1.47877610^{6}
```

Then we can check which hypothesis corresponds to the highest likelihood ratio, telling us which hypothesis we should believe in the most. To do this, enter:

```
> hypotheses[which.max(bfs)]
0.24
```

Now we know that a probability of 0.24 is our best guess, since this hypothesis produces the highest likelihood ratio when compared with the attendant's. In Chapter 10, you learned that using the mean or expectation of our data is often a good way to come up with a parameter estimate. Here we've simply chosen the hypothesis that individually explains the data the best, because we don't currently have a way to weigh our estimates by their probability of occurring.

Adding Priors to Our Likelihood Ratios

Now suppose you present your findings to the customer and the attendant. Both agree that your findings are pretty convincing, but then another person walks up to you and says, "I used to make games like these, and I can tell you that for some strange industry reason, the people who design these duck games never put the prize rate between 0.2 and 0.3. I'd bet you the odds are 1,000 to 1 that the real prize rate is not in this range. Other than that, I have no clue."

Now we have some prior odds that we'd like to use. Since the former game maker has given us some solid odds about his prior beliefs in the probability of getting a prize, we can try to multiply this by our current list of Bayes factors and compute the posterior odds. To do this, we create a list of prior odds ratios for every hypothesis we have. As the former game maker told us, the prior odds ratio for all probabilities between 0.2 and 0.3 should be 1/1,000. Since the maker has no opinion about other hypotheses, the odds ratio for these will just be 1. We can use a simple ifelse statement, using our vector of hypotheses, to create a vector of our odds ratios:

```
priors <- ifelse(hypotheses >= 0.2 & hypotheses <= 0.3, 1/1000,1)
```

Then we can once again use plot() to display this distribution of priors:

```
plot(hypotheses,priors,type='l')
```

Figure 19-2 shows our distribution of prior odds.

Because R is a vector-based language (for more information on this, see Appendix A), we can simply multiply our priors by our bfs and get a new vector of posteriors representing our Bayes factors:

```
posteriors  <- priors*bfs
```

Finally, we can plot a chart of the posterior odds of each of our many hypotheses:

```
plot(hypotheses,posteriors,type='l')
```

Figure 19-3 shows the plot.

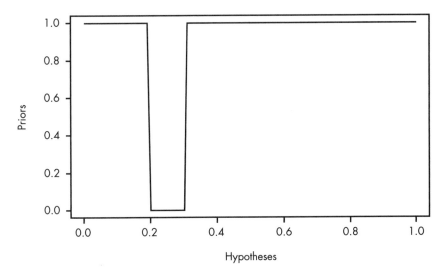

Figure 19-2: Visualizing our prior odds ratios

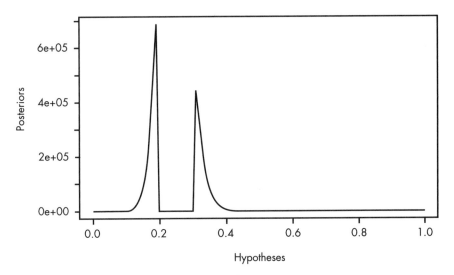

Figure 19-3: Plotting our distribution of Bayes factors

As we can see, we get a very strange distribution of possible beliefs. We have reasonable confidence in the values between 0.15 and 0.2 and between 0.3 and 0.35, but find the range between 0.2 and 0.3 to be extremely unlikely. But this distribution is an honest representation of the strength of belief in each hypothesis, given what we've learned about the duck game manufacturing process.

While this visualization is helpful, we really want to be able to treat this data like a true probability distribution. That way, we can ask questions about how much we believe in ranges of possible hypotheses and calculate the expectation of our distribution to get a single estimate for what we believe the hypothesis to be.

Building a Probability Distribution

A true probability distribution is one where the sum of all possible beliefs equals 1. Having a probability distribution would allow us to calculate the expectation (or mean) of our data to make a better estimate about the true rate of getting a prize. It would also allow us to easily sum ranges of values so we could come up with confidence intervals and other similar estimates.

The problem is that if we add up all the posterior odds for our hypotheses, they don't equal 1, as shown in this calculation:

```
> sum(posteriors)
3.140687510^{6}
```

This means we need to normalize our posterior odds so that they do sum to 1. To do so, we simply divide each value in our posteriors vector by the sum of all the values:

```
p.posteriors <- posteriors/sum(posteriors)
```

Now we can see that our p.posteriors values add up to 1:

```
> sum(p.posteriors)
1
```

Finally, let's plot our new p.posteriors:

```
plot(hypotheses,p.posteriors,type='l')
```

Figure 19-4 shows the plot.

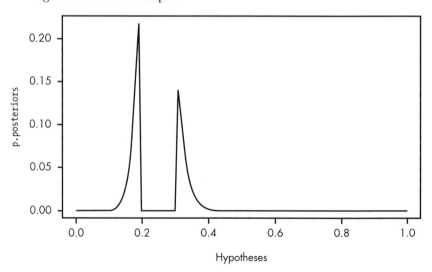

Figure 19-4: Our normalized posterior odds (note the scale on the y-axis)

We can also use our p.posteriors to answer some common questions we might have about our data. For example, we can now calculate the probability that the true rate of getting a prize is less than what the attendant claims. We just add up all the probabilities for values less than 0.5:

```
sum(p.posteriors[which(hypotheses < 0.5)])
> 0.9999995
```

As we can see, the probability that the prize rate is lower than the attendant's hypothesis is nearly 1. That is, we can be almost certain that the attendant is overstating the true prize rate.

We can also calculate the expectation of our distribution and use this result as our estimate for the true probability. Recall that the expectation is just the sum of the estimates weighted by their value:

```
> sum(p.posteriors*hypotheses)
0.2402704
```

Of course, we can see our distribution is a bit atypical, with a big gap in the middle, so we might want to simply choose the most *likely* estimate, as follows:

```
> hypotheses[which.max(p.posteriors)]
0.19
```

Now we've used the Bayes factor to come up with a range of probabilistic estimates for the true possible rate of winning a prize in the duck game. This means that we've used the Bayes factor as a form of parameter estimation!

From the Bayes Factor to Parameter Estimation

Let's take a moment to look at our likelihood ratios alone again. When we weren't using a prior probability for any of the hypotheses, you might have felt that we already had a perfectly good approach to solving this problem without needing the Bayes factor. We observed 24 ducks with prizes and 76 ducks without prizes. Couldn't we just use our good old beta distribution to solve this problem? As we've discussed many times since Chapter 5, if we want to estimate the rate of some event, we can always use the beta distribution. Figure 19-5 shows a plot of a beta distribution with an alpha of 24 and a beta of 76.

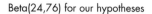

Beta(24,76) for our hypotheses

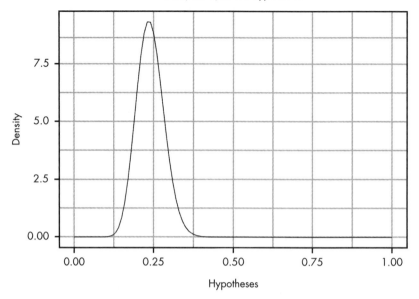

Figure 19-5: The beta distribution with an alpha of 24 and a beta of 76

Except for the scale of the y-axis, the plot looks nearly identical to the original plot of our likelihood ratios! In fact, if we do a few simple tricks, we can get these two plots to line up perfectly. If we scale our beta distribution by the size of our dx and normalize our bfs, we can see that these two distributions get quite close (Figure 19-6).

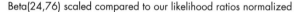

Beta(24,76) scaled compared to our likelihood ratios normalized

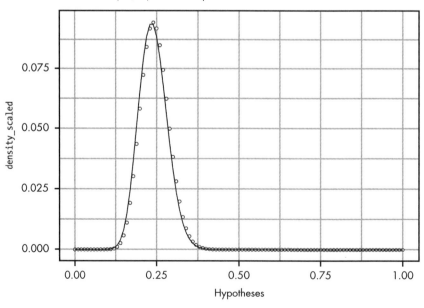

Figure 19-6: Our initial distribution of likelihood ratios maps pretty closely to Beta(24,76).

There seems to be only a slight difference now. We can fix it by using the weakest prior that indicates that getting a prize and not getting a prize are equally likely—that is, by adding 1 to both the alpha and beta parameters, as shown in Figure 19-7.

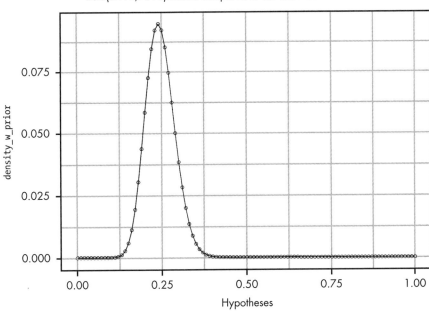

Figure 19-7: Our likelihood ratios map perfectly to a Beta(24+1,76+1) distribution.

Now we can see that the two distributions are perfectly aligned. Chapter 5 mentioned that the beta distribution was difficult to derive from our basic rules of probability. However, by using the Bayes factor, we've been able to empirically re-create a modified version of it that assumes a prior of Beta(1,1). And we did it without any fancy mathematics! All we had to do was:

1. Define the probability of the evidence given a hypothesis.
2. Consider all possible hypotheses.
3. Normalize these values to create a probability distribution.

Every time we've used the beta distribution in this book, we've used a beta-distributed prior. This made the math easier, since we can arrive at the posterior by combining the alpha and beta parameters from the likelihood and prior beta distributions. In other words:

$$\text{Beta}\left(\alpha_{\text{posterior}}, \beta_{\text{posterior}}\right) = \text{Beta}\left(\alpha_{\text{prior}} + \alpha_{\text{likelihood}}, \beta_{\text{prior}} + \beta_{\text{likelihood}}\right)$$

However, by building our distribution from the Bayes factor, we were able to easily use a unique prior distribution. Not only is the Bayes factor a great tool for setting up hypothesis tests, but, as it turns out, it's also all we need to create any probability distribution we might want to use to solve our problem, whether that's hypothesis testing or parameter estimation. We just need to be able to define the basic comparison between two hypotheses, and we're on our way.

When we built our A/B test in Chapter 15, we figured out how to reduce many hypothesis tests to a parameter estimation problem. Now you've seen how the most common form of hypothesis testing can also be used to perform parameter estimation. Given these two related insights, there is virtually no limit to the type of probability problems we can solve using only the most basic rules of probability.

Wrapping Up

Now that you've finished your journey into Bayesian statistics, you can appreciate the true beauty of what you've been learning. From the basic rules of probability, we can derive Bayes' theorem, which lets us convert evidence into a statement expressing the strength of our beliefs. From Bayes' theorem, we can derive the Bayes factor, a tool for comparing how well two hypotheses explain the data we've observed. By iterating through possible hypotheses and normalizing the results, we can use the Bayes factor to create a parameter estimate for an unknown value. This, in turn, allows us to perform countless other hypothesis tests by comparing our estimates. And all we need to do to unlock all this power is use the basic rules of probability to define our likelihood, $P(D \mid H)$!

Exercises

Try answering the following questions to see how well you understand using the Bayes factor and posterior odds to do parameter estimation. The solutions can be found at *https://nostarch.com/learnbayes/*.

1. Our Bayes factor assumed that we were looking at H_1: $P(\text{prize}) = 0.5$. This allowed us to derive a version of the beta distribution with an alpha of 1 and a beta of 1. Would it matter if we chose a different probability for H_1? Assume H_1: $P(\text{prize}) = 0.24$, then see if the resulting distribution, once normalized to sum to 1, is any different than the original hypothesis.

2. Write a prior for the distribution in which each hypothesis is 1.05 times more likely than the previous hypothesis (assume our dx remains the same).

3. Suppose you observed another duck game that included 34 ducks with prizes and 66 ducks without prizes. How would you set up a test to answer "What is the probability that you have a better chance of winning a prize in this game than in the game we used in our example?" Implementing this requires a bit more sophistication than the R used in this book, but see if you can learn this on your own to kick off your adventures in more advanced Bayesian statistics!

A

A QUICK INTRODUCTION TO R

In this book, we use the R programming language to do some tricky mathematical work for us. R is a programming language that specializes in statistics and data science. If you don't have experience with R, or with programming in general, don't worry—this appendix will get you started.

R and RStudio

To run the code examples in this book, you'll need to have R installed on your computer. To install R, visit *https://cran.rstudio.com/* and follow the installation steps for the operating system you're using.

Once you've installed R, you should also install *RStudio*, an integrated development environment (IDE) that makes it extremely easy to run R projects. Download and install RStudio from *www.rstudio.com/products/rstudio/download/*.

When you open RStudio, you should be greeted with several panels (Figure A-1).

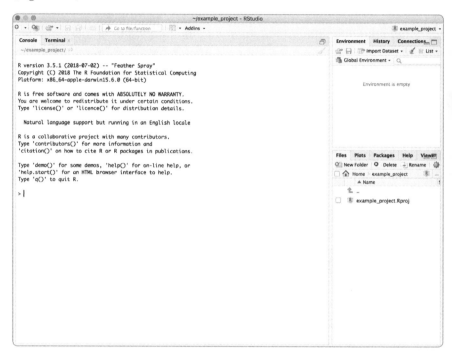

Figure A-1: Viewing the console in RStudio

The most important panel is the big one in the middle, called the *console*. In the console, you can enter any of the code examples from the book and run them simply by pressing ENTER. The console runs all the code you enter immediately, which makes it hard to keep track of the code you've written so far.

To write programs that you can save and come back to, you can place your code in an *R script*, which is a text file that you can load into the console later. R is an extremely interactive programming language, so rather than thinking of the console as a place you can test out code, think of R scripts as a way to quickly load tools you can use in the console.

Creating an R Script

To create an R script, go to **File▶New File▶R Script** in RStudio. This should create a new blank panel in the top left (Figure A-2).

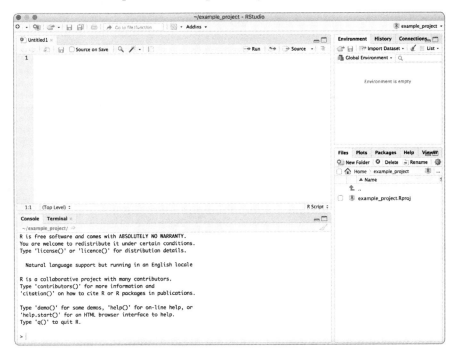

Figure A-2: Creating an R script

In this panel, you can enter code and save it as a file. To run the code, simply click the **Source** button at the top right of the panel, or run individual lines by clicking the **Run** button. The Source button will automatically load your file into the console as though you had typed it there yourself.

Basic Concepts in R

We'll be using R as an advanced calculator in this book, which means you'll only need to understand a few basics to work through the problems and extend the examples in the book on your own.

Data Types

All programming languages have different types of data, which you can use for different purposes and manipulate in different ways. R has a rich variety of types and data structures, but we'll only be using a very small number of them in this book.

Doubles

The numbers we use in R will all be of the type *double* (short for "double-precision floating-point," which is the most common way to represent decimal numbers on a computer). The double is the default type for representing decimal numbers. Unless otherwise specified, all numbers you enter into the console are of the double type.

We can manipulate numbers in the double type using standard mathematical operations. For example, we can add two numbers with the + operator. Try this out in the console:

```
> 5 + 2
[1] 7
```

We can also divide any numbers that give us decimal results using the / operator:

```
> 5/2
[1] 2.5
```

We can multiply values with the * operator like so:

```
> 5 * 2
[1] 10
```

and take the exponential of a value using the ^ operator. For example, 5^2 is:

```
> 5^2
[1] 25
```

We can also add - in front of a number to make it negative:

```
> 5 - -2
[1] 7
```

And we can also use scientific notation with e+. So 5×10^2 is just:

```
> 5e+2
[1] 500
```

If we use e- we get the same result as 5×10^{-2}:

```
> 5e-2
[1] 0.05
```

This is useful to know because sometimes R will return the result in scientific notation if it is too large to easily fit on the screen, like so:

```
> 5*10^20
[1] 5e+20
```

Strings

Another important type in R is the *string*, which is just a group of characters used to represent text. In R, we surround a string with quotation marks, like this:

```
> "hello"
[1] "hello"
```

Note that if you put a number inside a string, you can't use that number in regular numeric operations because strings and numbers are different types. For example:

```
> "2" + 2
Error in "2" + 2 : non-numeric argument to binary operator
```

We won't be making much use of strings in this book. We'll primarily use them to pass arguments to functions and to give labels to plots. But it's important to remember them if you're using text.

Logicals

Logical or *binary* types are true or false values represented by the codes TRUE and FALSE. Note that TRUE and FALSE aren't strings—they're not surrounded by quotes, and they're written in all uppercase. (R also allows you to simply use T or F instead of writing out the full words.)

We can combine logical types with the symbols & ("and") and | ("or") to perform basic logical operations. For example, if we wanted to know whether it's possible for something to be both true *and* false at the same time, we might enter:

```
> TRUE & FALSE
```

R would return:

```
[1] FALSE
```

telling us that a value can't be both true and false.

But what about true *or* false?

```
> TRUE | FALSE
[1] TRUE
```

Like strings, in this book logical values will primarily be used to provide arguments to functions we'll be using, or as the results of comparing two different values.

Missing Values

In practical statistics and data science, data is often missing some values. For example, say you have temperature data for the morning and afternoon of every day for a month, but something malfunctioned one day and you're missing a morning temperature. Because missing values are so common, R has a special way of representing them: using the value NA. It's important to have a way to handle missing values because they can mean very different things in different contexts. For example, when you're measuring rainfall a missing value might mean there was no rain in the gauge, or it might mean that there was plenty of rain but temperatures were freezing that night, cracking the gauge and causing all the water to leak out. In the first case, we might consider missing values to mean 0, but in the latter case it's not clear what the value should be. Keeping missing values separate from other values forces us to consider these differences.

To prompt us to make sense of what our missing values are whenever we try to use one, R will output NA for any operation using a missing value:

```
> NA + 2
[1] NA
```

As we'll see in a bit, various functions in R can handle missing values in different ways, but you shouldn't have to worry about missing values for the R you'll use in this book.

Vectors

Nearly every programming language contains certain features that make it unique and especially suited to solving problems in its domain. R's special feature is that it is a *vector language*. A vector is a list of values, and everything R does is an operation on a vector. We use the code c(...) to define vectors (but even if we put in just a single value, R does this for us!).

To understand how vectors work, let's consider an example. Enter the next example in a script, rather than the console. We first create a new vector by assigning the variable x to the vector c(1,2,3) using the assignment operator <- like so:

```
x <- c(1,2,3)
```

Now that we have a vector, we can use it in our calculations. When we perform a simple operation, like adding 3 to x, when we enter this in the console, we get a rather unexpected result (especially if you're used to another programming language):

```
> x + 3
[1] 4 5 6
```

The result of x + 3 tells us what happens if we add 3 to each value in our x vector. (In many other programming languages, we'd need to use a for loop or some other iterator to perform this operation.)

We can also add vectors to each other. Here, we'll create a new vector containing three elements, each with a value of 2. We'll name this vector y, then add y to x:

```
> y <- c(2,2,2)
> x + y
[1] 3 4 5
```

As you can see, this operation added each element in x to its corresponding element in y.

What if we multiply our two vectors?

```
> x * y
[1] 2 4 6
```

Each value in x was multiplied by its corresponding value in y. If the lists weren't the same size, or a multiple of the same size, then we'd get an error. If a vector is a multiple of the same size, R will just repeatedly apply the smaller vector to the larger one. However, we won't be making use of this feature in this book.

We can quite easily combine vectors in R by defining another vector based on the existing ones. Here, we'll create the vector z by combining x and y:

```
> z <- c(x,y)
> z
[1] 1 2 3 2 2 2
```

Notice that this operation didn't give us a vector of vectors; instead, we got a single vector that contains the values from both, in the order you set x and y when you defined z.

Learning to use vectors efficiently in R can be a bit tricky for beginners. Ironically, programmers who are experienced in a non-vector-based language often have the most difficulty. Don't worry, though: in this book, we'll use vectors to make reading code easier.

Functions

Functions are blocks of code that perform a particular operation on a value, and we'll use them in R to solve problems.

In R and RStudio, all functions come equipped with documentation. If you enter ? followed by a function name into the R console, you'll get the full documentation for that function. For example, if you enter ?sum into the RStudio console, you should see the documentation shown in Figure A-3 in the bottom-right screen.

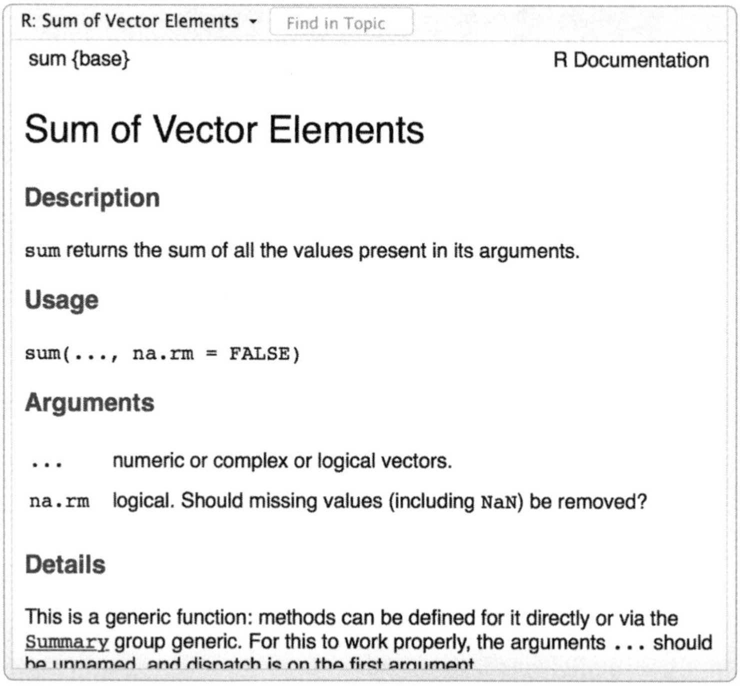

Figure A-3: Viewing the documentation for the sum() function

This documentation gives us the definition of the sum() function and some of its uses. The sum() function takes a vector's values and adds them all together. The documentation says it takes ... as an argument, which means it can accept any number of values. Usually these values will be a vector of numbers, but they can consist of multiple vectors, too.

The documentation also lists an *optional argument*: na.rm = FALSE. Optional arguments are arguments that you don't have to pass in to the function for it to work; if you don't pass an optional argument in, R will use the argument's default value. In the case of na.rm, which automatically removes any missing values, the default value, after the equal sign, is FALSE. That means that, by default, sum() won't remove missing values.

Basic Functions

Here are some of R's most important functions.

The length() and nchar() Functions

The length() function will return the length of a vector:

```
> length(c(1,2,3))
[1] 3
```

Since there are three elements in this vector, the length() function returns 3.

Because everything in R is a vector, you can use the length() function to find the length of anything—even a string, like "doggies":

```
> length("doggies")
[1] 1
```

R tells us that "doggies" is a vector containing one string.

Now, if we had two strings, "doggies" and "cats", we'd get:

```
> length(c("doggies","cats"))
[1] 2
```

To find the number of characters in a string, we use the nchar() function:

```
> nchar("doggies")
[1] 7
```

Note that if we use nchar() on the c("doggies","cats") vector, R returns a new vector containing the number of characters in each string:

```
> nchar(c("doggies","cats"))
[1] 7 4
```

The sum(), cumsum(), and diff() Functions

The sum() function takes a vector of numbers and adds all those numbers together:

```
> sum(c(1,1,1,1,1))
[1] 5
```

As we saw in the documentation in the previous section, sum() takes ... as its argument, which means it can accept any number of values:

```
> sum(2,3,1)
[1] 6
> sum(c(2,3),1)
[1] 6
> sum(c(2,3,1))
[1] 6
```

As you can see, no matter how many vectors we provide, sum() adds them up as though they were a single vector of integers. If you wanted to sum up multiple vectors, you'd call sum() on them each separately.

Remember, also, that the sum() function takes the optional argument na.rm, which by default is set to FALSE. The na.rm argument determines if sum() removes NA values or not.

If we leave na.rm set to FALSE, here's what happens if we try to use sum() on a vector with a missing value:

```
> sum(c(1,NA,3))
[1] NA
```

As we saw when NA was introduced, adding a value to an NA value results in NA. If we'd like R to give us a number as an answer instead, we can tell sum() to remove NA values by setting na.rm = TRUE:

```
> sum(c(1,NA,3),na.rm = TRUE)
[1] 4
```

The cumsum() function takes a vector and calculates its *cumulative sum*— a vector of the same length as the input that replaces each number with the sum of the numbers that come before it (including that number). Here's an example in code to make this clearer:

```
> cumsum(c(1,1,1,1,1))
[1] 1 2 3 4 5
> cumsum(c(2,10,20))
[1] 2 12 32
```

The diff() function takes a vector and subtracts each number from the number that precedes it in the vector:

```
> diff(c(1,2,3,4,5))
[1] 1 1 1 1
> diff(c(2,10,3))
[1]  8 -7
```

Notice that the result of the diff() function contains one fewer element than the original vector did. That's because nothing gets subtracted from the first value in the vector.

The : operator and the seq() Function

Often, rather than manually listing each element of a vector, we'd prefer to generate vectors automatically. To automatically create a vector of whole numbers in a certain range, we can use the : operator to separate the start and end of the range. R can even figure out if you want to count up or down (the c() wrapping this operator is not strictly necessary):

```
> c(1:5)
[1] 1 2 3 4 5

> c(5:1)
[1] 5 4 3 2 1
```

When you use :, R will count from the first value to the last.

Sometimes we'll want to count by something other than increments of one. The seq() function allows us to create vectors of a sequence of values that increment by a specified amount. The arguments to seq() are, in order:

1. The start of the sequence
2. The end of the sequence
3. The amount to increment the sequence by

Here are some examples of using seq():

```
> seq(1,1.1,0.05)
[1] 1.00 1.05 1.10

> seq(0,15,5)
[1]  0  5 10 15

> seq(1,2,0.3)
[1] 1.0 1.3 1.6 1.9
```

If we want to count down to a certain value using the seq() function, we use a minus value as our increment, like this:

```
> seq(10,5,-1)
[1] 10  9  8  7  6  5
```

The ifelse() Function

The ifelse() function tells R to take one of two actions based on some condition. This function can be a bit confusing if you're used to the normal if ... else control structure in other languages. In R, it takes the following three arguments (in order):

1. A statement about a vector that may be either true or false of its values
2. What happens in the case that the statement is true
3. What happens in the case that the statement is false

The ifelse() function operates on entire vectors at once. When it comes to vectors containing a single value, its use is pretty intuitive:

```
> ifelse(2 < 3,"small","too big")
[1] "small"
```

Here the statement is that 2 is smaller than 3, and we ask R to output "small" if it is, and "too big" if it isn't.

Suppose we have a vector x that contains multiple values:

```
> x <- c(1,2,3)
```

The ifelse() function will return a value for each element in the vector:

```
> ifelse(x < 3,"small","too big")
[1] "small"   "small"   "too big"
```

We can also use vectors in the results arguments for the ifelse(). Suppose that, in addition to our x vector, we had another vector, y:

```
y <- c(2,1,6)
```

We want to generate a new list that contains the greatest value from x and y for each element in the vector. We could use ifelse() to solve this very simply:

```
> ifelse(x > y,x,y)
[1] 2 2 6
```

You can see R has compared the values in x to the respective value in y and outputs the largest of the two for each element.

Random Sampling

We'll often use R to randomly sample values. This allows us to have the computer pick a random number or value for us. We use this sample to simulate activities like flipping a coin, playing "rock, paper, scissors," or picking a number between 1 and 100.

The runif() Function

One way to randomly sample values is with the function runif(), short for "random uniform," which takes a required argument n and gives that many samples in the range 0 to 1:

```
> runif(5)
[1] 0.8688236 0.1078877 0.6814762 0.9152730 0.8702736
```

We can use this function with ifelse() to generate a value A 20 percent of the time. In this case we'll use runif(5) to create five random values between 0 and 1. Then if the value is less than 0.2, we'll return "A"; otherwise, we'll return "B":

```
> ifelse(runif(5) < 0.2,"A","B")
[1] "B" "B" "B" "B" "A"
```

Since the numbers we're generating are random, we'll get a different result each time we run the ifelse() function. Here are some possible outcomes:

```
> ifelse(runif(5) < 0.2,"A","B")
[1] "B" "B" "B" "B" "B"
> ifelse(runif(5) < 0.2,"A","B")
 [1] "A" "A" "B" "B" "B"
```

The runif() function can take optional second and third arguments, which are the minimum and maximum values of the range to be uniformly sampled from. By default, the function uses the range between 0 and 1 inclusive, but you can set the range to be whatever you'd like:

```
> runif(5,0,2)
[1] 1.4875132 0.9368703 0.4759267 1.8924910 1.6925406
```

The rnorm() Function

We can also sample from a normal distribution using the rnorm() function, which we'll discuss in more depth in the book (the normal distribution is covered in Chapter 12):

```
> rnorm(3)
[1]  0.28352476  0.03482336 -0.20195303
```

By default, rnorm() samples from a normal distribution with a mean of 0 and standard deviation of 1, as is the case in this example. For readers unfamiliar with the normal distribution, this means that samples will have a "bell-shaped" distribution around 0, with most samples being close to 0 and very few being less than −3 or greater than 3.

The rnorm() function has two optional arguments, mean and sd, which allow you to set a different mean and standard deviation, respectively:

```
> rnorm(4,mean=2,sd=10)
[1] -12.801407  -9.648737   1.707625  -8.232063
```

In statistics, sampling from a normal distribution is often more common than sampling from a uniform distribution, so rnorm() comes in quite handy.

The sample() Function

Sometimes, we want to sample from something other than just a well-studied distribution. Suppose you have a drawer containing socks of many colors:

```
socks <- c("red","grey","white","red","black")
```

If you wanted to simulate the act of randomly picking any two socks, you could use R's `sample()` function, which takes as arguments a vector of values and the number of elements to sample:

```
> sample(socks,2)
[1] "grey" "red"
```

The `sample()` function behaves as though we've picked two random socks out of the drawer—without putting any back. If we sample five socks, we'll get all of the socks we originally had in the drawer:

```
> sample(socks,5)
[1] "grey" "red"   "red"   "black" "white"
```

That means that if we try to take six socks from the drawer where there are only five available socks, we'll get an error:

```
> sample(socks,6)
Error in sample.int(length(x), size, replace, prob) :
  cannot take a sample larger than the population when 'replace = FALSE'
```

If we want to both sample and "put the socks back," we can set the optional argument `replace` to `TRUE`. Now, each time we sample a sock, we put it back in the drawer. This allows us to sample more socks than are in the drawer. It also means the distribution of socks in the drawer never changes.

```
> sample(socks,6,replace=TRUE)
[1] "black" "red"   "black" "red"   "black" "black"
```

With these simple sampling tools, you can run surprisingly sophisticated simulations in R that save you from doing a lot of math.

Using set.seed() for Predictable Random Results

The "random numbers" generated by R aren't truly random numbers. As in all programming languages, random numbers are generated by a *pseudorandom number generator*, which takes a *seed value* and uses that to create a sequence of numbers that are random enough for most purposes. The seed value sets the initial state of the random number generator and determines which numbers will come next in the sequence. In R, we can manually set this seed using the `set.seed()` function. Setting the seed is extremely useful for cases when we want to use the same random results again:

```
> set.seed(1337)
> ifelse(runif(5) < 0.2,"A","B")
[1] "B" "B" "A" "B" "B"
```

```
> set.seed(1337)
> ifelse(runif(5) < 0.2,"A","B")
[1] "B" "B" "A" "B" "B"
```

As you can see, when we used the same seed twice with the runif() function, it generated the same set of supposedly random values. The main benefit of using set.seed() is making the results reproducible. This can make tracking down bugs in programs that involve sampling much easier, since the results don't change each time the program is run.

Defining Your Own Functions

Sometimes it's helpful to write our own functions for specific operations we'll have to perform repeatedly. In R, we can define functions using the keyword function (a *keyword* in a programming language is simply a special word reserved by the programming language for a specific use).

Here's the definition of a function that takes a single argument, val—which here stands for the value the user will input to the function—and then doubles val and cubes it.

```
double_then_cube <- function(val){
  (val*2)^3
}
```

Once we've defined our function, we can use it, just like R's built-in functions. Here's our double_then_cube() function applied to the number 8:

```
> double_then_cube(8)
[1] 4096
```

Also, because everything we did to define our function is *vectorized* (that is, all values work on vectors of values), our function will work on vectors as well as single values:

```
> double_then_cube(c(1,2,3))
[1] 8 64 216
```

We can define functions that take more than one argument as well. The sum_then_square() function, defined here, adds two arguments together, then squares the result:

```
sum_then_square <- function(x,y){
  (x+y)^2
}
```

By including the two arguments (x,y) in the function definition, we're telling R the sum_then_square() function expects two arguments. Now we can use our new function, like this:

```
> sum_then_square(2,3)
[1] 25
> sum_then_square(c(1,2),c(5,3))
[1] 36 25
```

We can also define functions that require multiple lines. In R, when a function is called it will always return the result of the calculation on the final line of the function definition. That means we could have rewritten sum_then_square() like this:

```
sum_then_square <- function(x,y){
  sum_of_args <- x+y
  square_of_result <- sum_of_args^2
  square_of_result
}
```

Typically, when you write functions, you'll want to write them in an R script file so you can save them and reuse them later.

Creating Basic Plots

In R, we can quickly generate plots of data very easily. Though R has an extraordinary plotting library called ggplot2, which contains many useful functions for generating beautiful plots, we'll restrict ourselves to R's base plotting functions for now, which are plenty useful on their own.

To show how plotting works, we'll create two vectors of values, our xs and our ys:

```
> xs <- c(1,2,3,4,5)
> ys <- c(2,3,2,4,6)
```

Next, we can use these vectors as arguments to the plot() function, which will plot our data for us. The plot() function takes two arguments: the values of the plot's points on the x-axis and the values of those points on the y-axis, in that order:

```
> plot(xs,ys)
```

This function should generate the plot shown in Figure A-4 in the bottom-left window of RStudio.

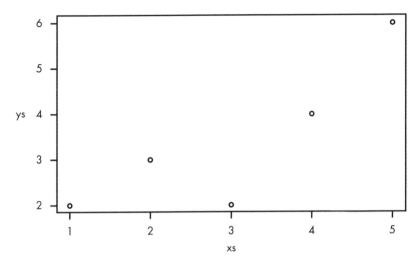

Figure A-4: A simple plot created with R's plot() function

This plot shows the relationship between our xs values and their corresponding ys values. If we return to the function, we can give this plot a title using the optional main argument. We can also change the x- and y-axis labels with the xlab and ylab arguments, like this:

```
plot(xs,ys,
      main="example plot",
      xlab="x values",
      ylab="y values"
      )
```

The new labels should show up as they appear in Figure A-5.

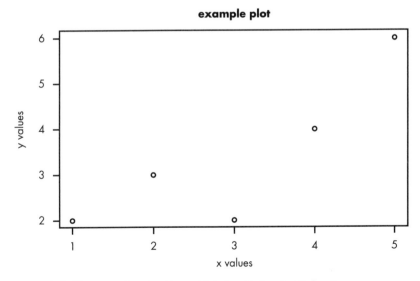

Figure A-5: Changing the plot title and labels with the plot() function

We can also change the plot's type using the type argument. The first kind of plot we generated is called a *point plot*, but if we wanted to make a line plot, which draws a line through each value, we could set type="l":

```
plot(xs,ys,
     type="l",
     main="example plot",
     xlab="x values",
     ylab="y values"
     )
```

It would then look like Figure A-6.

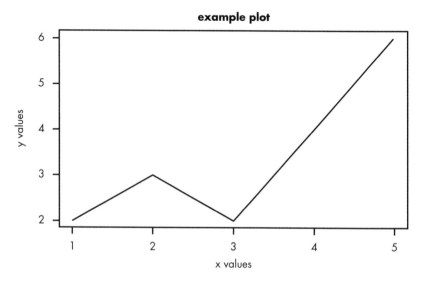

Figure A-6: A line plot generated with R's plot() function

Or we can do both! An R function called lines() can add lines to an existing plot. It takes most of the same arguments as plot():

```
plot(xs,ys,
     main="example plot",
     xlab="x values",
     ylab="y values"
     )
lines(xs,ys)
```

Figure A-7 shows the plot this function would generate.

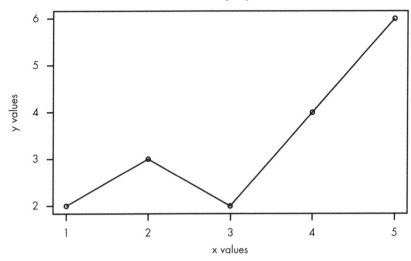

Figure A-7: Adding lines to an existing plot with R's lines() function

There are many more amazing ways to use R's basic plots, and you can consult ?plot for more information on them. However, if you want to create truly beautiful plots in R, you should research the ggplot2 library (*https:// ggplot2.tidyverse.org/*).

Exercise: Simulating a Stock Price

Now let's put everything we've learned together to create a simulated stock ticker! People often model stock prices using the cumulative sum of normally distributed random values. To start, we'll simulate stock movement for a period of time by generating a sequence of values from 1 to 20, incrementing by 1 each time using the seq() function. We'll call the vector representing the period of time t.vals.

```
t.vals <- seq(1,20,by=1)
```

Now t.vals is a vector containing the sequence of numbers from 1 to 20 incremented by 1. Next, we'll create some simulated prices by taking the cumulative sum of a normally distributed value for each time in your t.vals. To do this we'll use rnorm() to sample the number of values equal to the length of t.vals. Then we'll use cumsum() to calculate the cumulative sum of this vector of values. This will represent the idea of a price moving up or down based on random motion; less extreme movements are more common than more extreme ones.

```
price.vals <- cumsum(rnorm(length(t.vals),mean=5,sd=10))
```

Finally, we can plot all these values to see how they look! We'll use both the plot() and lines() functions, and label the axes according to what they represent.

```
plot(t.vals,price.vals,
    main="Simulated stock ticker",
    xlab="time",
    ylab="price")
lines(t.vals,price.vals)
```

The plot() and lines() functions should generate the plot shown in Figure A-8.

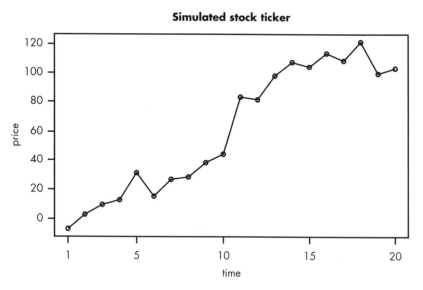

Figure A-8: The plot generated for our simulated stock ticker

Summary

This appendix should cover enough R to give you a grasp of the examples in this book. I recommend following along with the book's chapters, then playing around by modifying the code examples to learn more. R also has some great online documentation if you want to take your experimentation further.

B

ENOUGH CALCULUS TO GET BY

In this book, we'll occasionally use ideas from calculus, though no actual manual solving of calculus problems will be required! What *will* be required is an understanding of some of the basics of calculus, such as the derivative and (especially) the integral. This appendix is by no means an attempt to teach these concepts deeply or show you how to solve them; instead, it offers a brief overview of these ideas and how they're represented in mathematical notation.

Functions

A *function* is just a mathematical "machine" that takes one value, does something with it, and returns another value. This is very similar to how functions in R work (see Appendix A): they take in a value and return a result. For example, in calculus we might have a function called f defined like this:

$$f(x) = x^2$$

In this example, f takes a value, x, and squares it. If we input the value 3 into f, for example, we get:

$$f(3) = 9$$

This is a little different than how you might have seen it in high school algebra, where you'd usually have a value y and some equation involving x.

$$y = x^2$$

One reason why functions are important is that they allow us to abstract away the actual calculations we're doing. That means we can say something like $y = f(x)$, and just concern ourselves with the abstract behavior of the function itself, not necessarily how it's defined. That's the approach we'll take for this appendix.

As an example, say you're training to run a 5 km race and you're using a smartwatch to keep track of your distance, speed, time, and other factors. You went out for a run today and ran for half an hour. However, your smartwatch malfunctioned and recorded only your speed in miles per hour (mph) throughout your 30-minute run. Figure B-1 shows the data you were able to recover.

For this appendix, think of your running speed as being created by a function, s, that takes an argument t, the time in hours. A function is typically written in terms of the argument it takes, so we would write $s(t)$, which results in a value that gives your current speed at time t. You can think of the function s as a machine that takes the current time and returns your speed at that time. In calculus, we'd usually have a specific definition of $s(t)$, such as $s(t) = t^2 + 3t + 2$, but here we're just talking about general concepts, so we won't worry about the exact definition of s.

NOTE *Throughout the book we'll be using R to handle all our calculus needs, so it's really only important that you understand the fundamental ideas behind it, rather than the mechanics of solving calculus problems.*

From this function alone, we can learn a few things. It's clear that your pace was a little uneven during this run, going up and down from a high of nearly 8 mph near the end and a low of just under 4.5 mph in the beginning.

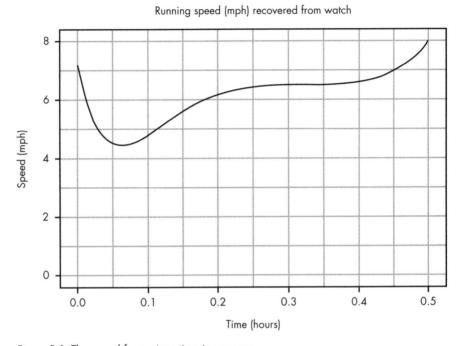

Figure B-1: The speed for a given time in your run

However, there are still a lot of interesting questions you might want to answer, such as:

- How far did you run?
- When did you lose the most speed?
- When did you gain the most speed?
- During what times was your speed relatively consistent?

We can make a fairly accurate estimate of the last question from this plot, but the others seem impossible to answer from what we have. However, it turns out that we can answer *all* of these questions with the power of calculus! Let's see how.

Determining How Far You've Run

So far our chart just shows your running speed at a certain time, so how do we find out how far you've run?

This doesn't sound too difficult in theory. Suppose, for example, you ran 5 mph consistently for the whole run. In that case, you ran 5 mph for 0.5 hour, so your total distance was 2.5 miles. This intuitively makes sense, since you would have run 5 miles each hour, but you ran for only half an hour, so you ran half the distance you would have run in an hour.

But our problem involves a different speed at nearly every moment that you were running. Let's look at the problem another way. Figure B-2 shows the plotted data for a constant running speed.

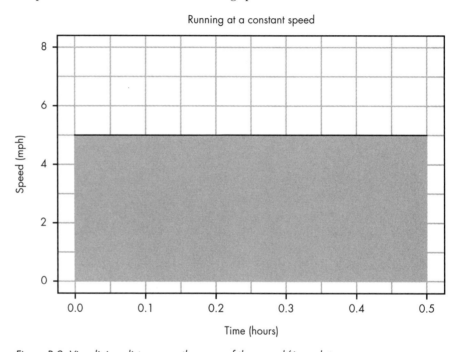

Figure B-2: Visualizing distance as the area of the speed/time plot

You can see that this data creates a straight line. If we think about the space under this line, we can see that it's a big block that actually represents the distance you've run! The block is 5 high and 0.5 long, so the area of this block is 5 × 0.5 = 2.5, which gives us the 2.5 miles result!

Now let's look at a simplified problem with varying speeds, where you ran 4.5 mph from 0.0 to 0.3 hours, 6 mph from 0.3 to 0.4 hours, and 3 mph the rest of the way to 0.5 miles. If we visualize these results as blocks, or *towers*, as in Figure B-3, we can solve our problem the same way.

The first tower is 4.5 × 0.3, the second is 6 × 0.1, and the third is 3 × 0.1, so that:

$$4.5 \times 0.3 + 6 \times 0.1 + 3 \times 0.1 = 2.25$$

By looking at the area under the tower, then, we get the total distance you traveled: 2.25 miles.

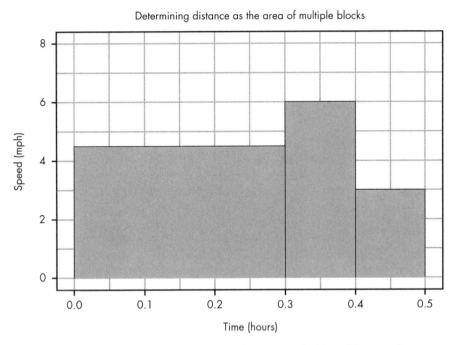

Figure B-3: We can easily calculate your total distance traveled by adding together these towers.

Measuring the Area Under the Curve: The Integral

You've now seen that we can figure out the area under the line to tell us how far you traveled. Unfortunately, the line for our original data is curved, which makes our problem a bit difficult: how can we calculate the towers under our curvy line?

We can start this process by imagining some large towers that are fairly close to the pattern of our curve. If we start with just three towers, as we can see in Figure B-4, it isn't a bad estimate.

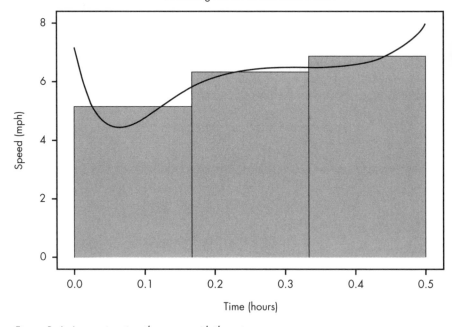

Figure B-4: Approximating the curve with three towers

By calculating the area under each of these towers, we get a value of 3.055 miles for your estimated total miles traveled. But we could clearly do better by making more, smaller towers, as shown in Figure B-5.

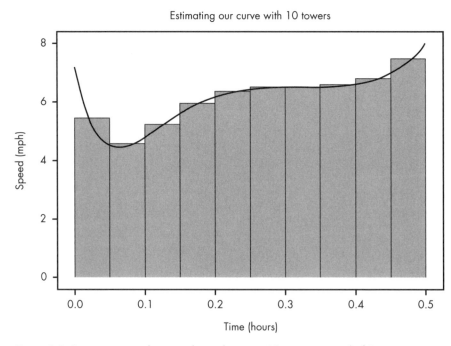

Figure B-5: Approximating the curve better by using 10 towers instead of 3

Adding up the areas of these towers, we get 3.054 miles, which is a more accurate estimate.

If we imagine repeating this process forever, using more and thinner towers, eventually we would get the full area under the curve, as in Figure B-6.

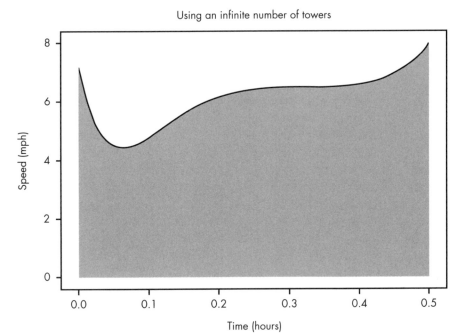

Figure B-6: Completely capturing the area under the curve

This represents the exact area traveled for your half-hour run. If we could add up infinitely many towers, we would get a total of 3.053 miles. Our estimates were pretty close, and as we use more and smaller towers, our estimate gets closer. The power of calculus is that it allows us to calculate this *exact* area under the curve, or the *integral*. In calculus, we'd represent the integral for our $s(t)$ from 0 to 0.5 in mathematical notation as:

$$\int_0^{0.5} s(t)\,dt$$

That $\int$ is just a fancy S, meaning the sum (or total) of the area of all the little towers in $s(t)$. The dt notation reminds us that we're talking about little bits of the variable t; the d is a mathematical way to refer to these little towers. Of course, in this bit of notation, there's only one variable, t, so we aren't likely to get confused. Likewise, in this book, we typically drop the dt (or its equivalent for the variable being used) since it's obvious in the examples.

In our last notation we set the beginning and end of our integral, which means we can find the distance not just for the whole run but also for a section of it. Suppose we wanted to know how far you ran between 0.1 to 0.2 of an hour. We would note this as:

$$\int_{0.1}^{0.2} s(t)\,dt$$

We can visualize this integral as shown in Figure B-7.

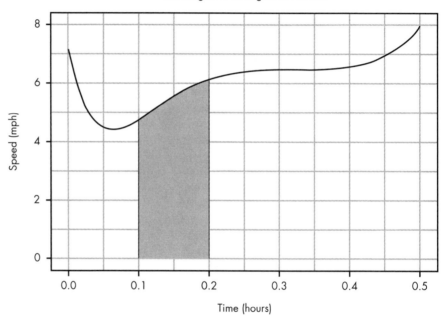

Figure B-7: Visualizing the area under the curve for the region from 0.1 to 0.2

The area of just this shaded region is 0.556 miles.

We can even think of the integral of our function as another function. Suppose we define a new function, dist(T), where T is our "total time run":

$$\text{dist}(T) = \int_{0}^{T} s(t)\,dt$$

This gives us a function that tells us the *distance* you've traveled at time T. We can also see why we want to use *dt* because we can see that our integral is being applied to the lowercase *t* argument rather than the capital *T* argument. Figure B-8 plots this out to the total distance you've run at any given time *T* during your run.

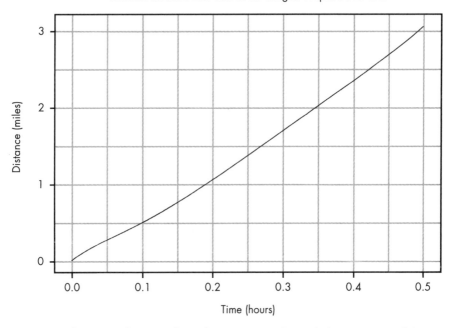

Distance traveled over time as the integral of speed over time

Figure B-8: Plotting out the integral transforms a time and speed plot to a time and distance plot.

In this way, the integral has transformed our function *s*, which was "speed at a time," to a function *dist*, "distance covered at a time." As shown earlier, the integral of our function between two points represents the distance traveled between two different times. Now we're looking at the total distance traveled at any given time *t* from the beginning time of 0.

The integral is important because it allows us to calculate the area under curves, which is much trickier to calculate than if we have straight lines. In this book, we'll use the concept of the integral to determine the probabilities that events are between two ranges of values.

Measuring the Rate of Change: The Derivative

You've seen how we can use the integral to figure out the distance traveled when all we have is a recording of your speed at various times. But with our varying speed measurements, we might also be interested in figuring out the *rate of change* for your speed at various times. When we talk about the rate at which speed is changing, we're referring to *acceleration*. In our chart, there are a few interesting points regarding the rate of change: the points when you're losing speed the fastest, when you're gaining speed the fastest, and when the speed is the most steady (i.e., the rate of change is near 0).

Just as with integration, the main challenge of figuring out your acceleration is that it seems to always be changing. If we had a constant rate of change,
calculating the acceleration isn't that difficult, as shown in Figure B-9.

A constant rate of increase in speed

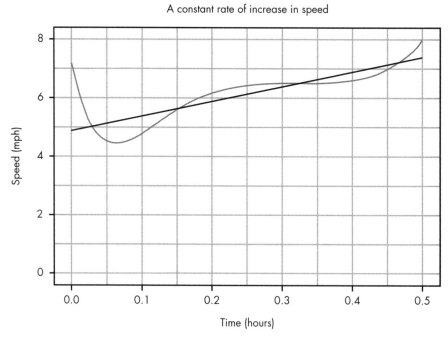

Figure B-9: Visualizing a constant rate of change (compared with your actual
changing rate)

You might remember from basic algebra that we can draw any line
using this formula:

$$y = mx + b$$

where b is the point at which the line crosses the y-axis and m is the slope of
the line. The *slope* represents the rate of change of a straight line. For the
line in Figure B-9, the full formula is:

$$y = 5x + 4.8$$

The slope of 5 means that for every time x grows by 1, y grows by 5; 4.8
is the point at which the line crosses the x-axis. In this example, we'd interpret this formula as $s(t) = 5t + 4.8$, meaning that for every mile you travel

you accelerate by 5 mph, and that you started off at 4.8 mph. Since you've run half a mile, using this simple formula, we can figure out:

$$s(t) = 5 \times 0.5 + 4.8 = 7.3$$

which means at the end of your run, you would be traveling 7.3 mph. We could similarly determine your exact speed at any point in the run, as long as the acceleration is constant!

For our actual data, because the line is curvy it's not easy to determine the slope at a single point in time. Instead, we can figure out the slopes of parts of the line. If we divide our data into three subsections, we could draw lines between each part as in Figure B-10.

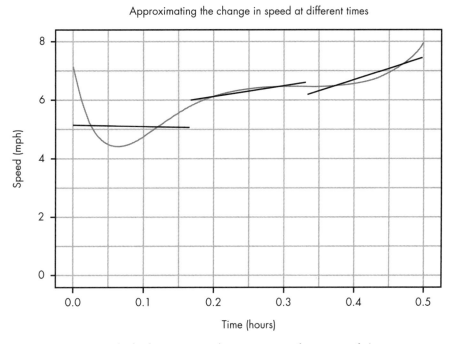

Figure B-10: Using multiple slopes to get a better estimate of your rate of change

Now, clearly these lines aren't a perfect fit to our curvy line, but they allow us to see the parts where you accelerated the fastest, slowed down the most, and were relatively stable.

If we split our function up into even more pieces we can get even better estimates, as in Figure B-11.

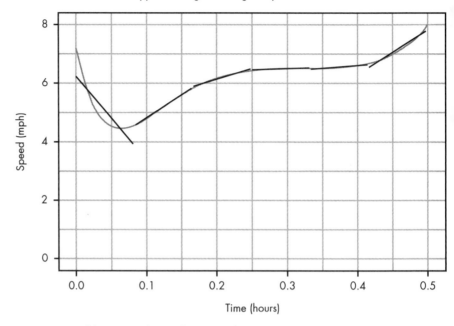

Figure B-11: Adding more slopes allows us to better approximate your curve.

Here we have a similar pattern to when we found the integral, where we split the area under the curve into smaller and smaller towers until we were adding up infinitely many small towers. Now we want to break up our line into infinitely many small line segments. Eventually, rather than a single m representing our slope, we have a new function representing the rate of change at each point in our original function. This is called the *derivative*, represented in mathematical notation like this:

$$\frac{d}{dx} f(x)$$

Again, the dx just reminds us that we're looking at very small pieces of our argument x. Figure B-12 shows the plot of the derivative for our $s(t)$ function, which allows us to see the exact rate of speed change at each moment in your run. In other words, this is a plot of your acceleration during your run. Looking at the y-axis, you can see that you rapidly lost speed in the beginning, and at around 0.3 hours you had a period of 0 acceleration, meaning your pace did not change (this is usually a good thing when practicing for a race!). We can also see exactly when you gained the most speed. Looking at the original plot, we couldn't easily tell if you were gaining speed faster around 0.1 hours (just after your first speedup) or at the end of your run. With the derivative, though, it's clear that the final burst of speed at the end was indeed faster than at the beginning.

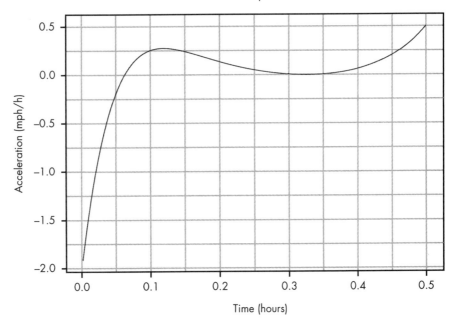

The derivative of speed: acceleration

Figure B-12: The derivative is another function that describes the slope of s(x) at each point.

The derivative works just like the slope of a straight line, only it tells us how much a curvy line is sloping at a certain point.

The Fundamental Theorem of Calculus

We'll look at one last truly remarkable calculus concept. There's a very interesting relationship between the integral and the derivative. (Proving this relationship is far beyond the scope of this book, so we'll focus only on the relationship itself here.) Suppose we have a function $F(x)$, with a capital F. What makes this function special is that *its derivative* is $f(x)$. For example, the derivative of our dist function is our s function; that is, your change in distance at each point in time is your speed. The derivative of speed is acceleration. We can describe this mathematically as:

$$\frac{d}{dx}F(x) = f(x)$$

In calculus terms we call F the *antiderivative* of f, because f is F's derivative. Given our examples, the antiderivative of acceleration would be speed, and the antiderivative of speed would be distance. Now suppose for any value of f, we want to take its integral between 10 and 50; that is, we want:

$$\int_{10}^{50} f(x)\,dx$$

We can get this simply by subtracting $F(10)$ from $F(50)$, so that:

$$\int_{10}^{50} f(x)\,dx = F(50) - F(10)$$

The relationship between the integral and the derivative is called the *fundamental theorem of calculus*. It's a pretty amazing tool, because it allows us to solve integrals mathematically, which is often much more difficult than finding derivatives. Using the fundamental theorem, if we can find the antiderivative of the function we want to find the integral of, we can easily perform integration. Figuring this out is the heart of performing integration by hand.

A full course on calculus (or two) typically explores the topics of integrals and derivatives in much greater depth. However, as mentioned, in this book we'll only be making occasional use of calculus, and we'll be using R for all of the calculations. Still, it's helpful to have a rough understanding of what calculus and those unfamiliar $\int$ symbols are all about!

INDEX

Bayes' theorem (*continued*)
 evidence observation, use in, 65
 formula, 64, 67, 158
 LEGO example (*see* LEGO
  visualization)
 likelihood, 158 (*see also* likelihood)
 posterior probabilities, 158 (*see also*
  posterior probabilities)
 prior probability, 158 (*see also* prior
  probabilities)
 proportional form, 87, 158
 statistics, importance to, 64, 67
beliefs
 data, relationship between, 10–11,
  64–65, 74
 distribution of, 84–85
 irrational, 179
 measuring, 18–19
 mutability of, 11
 origins of, 11
 prior, 5, 86, 141, 143–144, 157,
  170–171
 probability distribution of, 49, 83
 ranges of, 83
 strength of, analyzing, 64–65, 67,
  80–81
 worldview, relationship between,
  64–65
beta distributions
 applications, 45
 beliefs, of, 87
 binomial distributions, *versus*, 50, 52
 changes in, as information is
  gained, 138, 142
 estimating with, 121
 Gacha game example, use in, 54
 mean of, 125
 normal distributions, compared to,
  121–122
 normalizing values, 51
 overview, 45
 parameters, 84, 135, 140
 prize distribution example, 191–194
 probability density function of,
  51–52, 124
 probability, true, 121
 quantiles, of, 135
 undefined, 145
binomial coefficients, 36–38. *See also*
  combinatorics

binomial distributions
 beta distributions, *versus*, 50, 52
 examples, 34
 outcomes, 34
 overview, 34
 parameters, 34
 Probability Mass Function (PMF),
  relationship between, 39
 probability, use in calulating, 47–48
 shorthand notations, 34–35
 solving for, 43
 structure, 34–35

C

C-3PO example, 84–88
c(), 200
calculus, fundamental theorem of, 228
change, rate of, 223
click-through rate example, 149
 A/B test
  conversion rate, 150–152
  data collection, 151
  parameters, 151–152
  prior probability, finding, 150
  setting up the A/B test, 150
 Monte Carlo simulations, 152–153,
  153–154, 154–155
combinatorics, 16, 37–38
conditional probabilities, 5–6
 beliefs, impact of, 6
 color blindness example, 61–64
 defining, 60, 65
 experiences, impact of, 6
 flu vaccine risks example, 60–61
 likelihoods, 64
 overview, 60
 reversing, 62–63, 67–68
confidence intervals (CIs), 132–133
conspiracy theories, 181
continuous distributions, quantifying,
  52–53, 55
conversion event, estimating. *See also*
  click-through rate example
 beta distributions for, 138, 139–140
 PDF, using, 124–127, 135
 prior probabilities regarding,
  141–142
conversion rate, 124
critical region, 132–133
cumsum(), 204

cumulative distribution function (CDF)
    antiderivative of a PDF, 127
    confidence intervals, estimating,
        132–133
    distribution, sums of the parts of, 127
    interpreting, 130
    intuitiveness, 128
    inverse of, 133–134
    mean of, 131
    median of, 130, 133
    quantile function, use in
        calculating, 133–134
    R programming language, use
        in, 132
    usefulness, 132
    visualization of, 128, 130

D

data
    Bayesian statistics, importance to, 63
    beliefs, relationship between,
        10, 11, 64–65, 74
    high-probability, 7
    hypothesis, relationship between,
        6–10
    normalizing, 76–77
    observation of, 4, 7
    probability, relationship between,
        10, 76, 77
    size of sets, 75
    spread, measuring, 103–105, 108
dbeta(), 53, 126–127
derivative, 223–227
dfunction(), 126
diff(), 204
dnorm(), 126, 133

E

errors, statistical analysis, 84
error value, 105
evidence, observation of, 65
expectation, defining, 99. *See also* mean
exponential penalty, 107

F

factorials, 37
falsifiability, 180–181
Frequentist Fighters! game example,
        53–54

function, 209
functions
    calculus, use of, 216–217, 218–220
    defining, 216
    integrals, use of (*see* integrals)
fundamental theorem of calculus, 228

G

Gacha games
    Bayesian Battlers game example,
        41–43
    Frequentist Fighters! game example,
        53–54
    probability distributions, 41–43
    reverse-engineering, 53–54
ggplot2, 210

H

Han Solo example, 84, 86, 139
hypotheses
    alternate, 9–10, 11, 78–79, 178–179
    beliefs, relationship between, 74, 83
    confidence in, 83
    developing, 48
    formal, 7
    formulating, 4, 6–7
    hidden, 180
    infinite, 49–50, 55
    multiple, 183, 185
    probability, relationship between,
        7, 17–18, 47, 48–49
    R programming language,
        searching with (*see* R
        progamming language)
    testing, 49

I

ifelse, 188, 205–206
independent probabilities, 59, 61
inference, 46–47, 48
infinity, 117
integrals, 51, 119
    applications, 53
    approximating, visually, 131–132
    beta distribution of, 128
    derivative, relationship between, 228
    estimating with, 219–223
    using R to solve for, 131
integrate() function, 53, 118
intuition, use of, 70–72, 107, 108, 128

R

R programming language
 CDF, use of in, 132
 doubles, 198
 functions, defining, 209–210
 installing, 196
 integrals, solving for, 131
 likelihood ratios, calulating, 188–189
 logicals, 199
 missing values, 200
 multiple hypotheses testing with,
  186–188
 plots, generating, 210–213
 probability density dunction (PDF),
  use of, 126–127
 quantiles, calculating, 135
 random numbers generated by,
  208–209
 R Script, 197
 strings, 199, 203
 vectors, 200–202
  length(), 202–203
  sum(), 203
rate of change, 223
ratio, coin toss, 19
ratio of posteriors, 158
 posterior odds, 159, 170–171, 177
reasoning, 70, 72, 73. *See also* Bayesian
  reasoning
reciprocal, multiplying, 106
rnorm, 207
RStudio, 196
runif(), 206–207

S

sample(), 207–208
seq(), 186, 204–205
set.seed(), 208–209
sigma events, 120
skepticism, 175–176
spatial reasoning, 73
spread, 103–106, 108
standard deviation, 103
 calculating, 112
 determining, 107–108
 formula, 108

parameter in normal distribution,
  111, 114
 usefulness, 108–109, 111
statistical analysis, 84–85
statistical reasoning, 8, 85
statistics. *See also* inference
 conditional probabilities,
  importance of, 60 (*see also*
  conditional probabilities)
 probability, *versus,* 46
stock price simulation, 213–214
sum rule for non-mutually exclusive
  events, 28–30, 61
sum_then_square(), 209–210
summation symbol, 42

T

true value, 105
Twilight Zone example, 168-172,
  175–176

U

uncertainty, measuring, 13–14

V

variance, 103
 finding, 106–107
 properties, 107
 squaring, 107, 109
vectors
 length(), 202–203
 plots, as part of, 210
 R, 200–202
 stock price simulation, 213–214
 sum(), 203
villain bomb example, 112–114,
  116–118, 120

W

weighted sums, 98